Machine Learning at Scale with H2O

A practical guide to building and deploying machine learning models on enterprise systems

Gregory Keys

David Whiting

BIRMINGHAM—MUMBAI

Machine Learning at Scale with H2O

Copyright © 2022 Packt Publishing

Publishing Product Manager: Aditi Gour

Senior Editor: David Sugarman

Content Development Editor: Manikandan Kurup

Technical Editor: Rahul Limbachiya

Copy Editor: Safis Editing

Project Coordinator: Farheen Fathima

Proofreader: Safis Editing

Indexer: Subalakshmi Govindhan

Production Designer: Alishon Mendonca

Marketing Coordinator: Abeer Riyaz Dawe

First published: July 2022

Production reference: 1290622

Published by Packt Publishing Ltd.

Livery Place

35 Livery Street

Birmingham

B3 2PB, UK.

ISBN 978-1-80056-601-9

www.packt.com

My deepest love and warmth to Mary, Julia and Alexa for their support and understanding while husband and dad disappeared to the basement for significant chunks of nights and weekends as the seasons progressed.

- Gregory

To my wife Kathy, and son Ben, who endured too many late nights and weekends of dad locked away in his study working; the book has been a family effort and its culmination a family success.

- David

Acknowledgments

This book would not have been possible without the approval and support of our respective leaders at H2O.ai at the time of its writing, Dmitry Baev and Eyal Kaldes. In addition, we pay our great appreciation to the deep expertise of the many *Makers* at H2O. ai. Their day-to-day collaboration, education, and machine learning expertise are diffused throughout the pages of this book.

One name needs to be called out in particular: massive thanks to Eric Gudgeon for his infinite and unrelenting technical teachings, and for defining and developing a vast landscape of H2O model deployment implementations.

This book took longer to pull together than either of us expected. Working at a hyper-focused and highly energized company certainly was a contributing factor. Against this backdrop, we appreciate the world-class patience, encouragement, guidance, and professionalism of the Packt team in collaborating on this book from start to finish.

And most importantly there is family, who unfairly signed up for book writing without fully knowing it.

Contributors

About the authors

Gregory Keys is a master principal cloud architect for Data and AI at Oracle. Formerly a senior solutions architect at H2O.ai, he has over 20 years of experience designing and implementing software and data systems. He specializes in AI/ML solutions and has multiple software patents. Gregory has a PhD in evolutionary biology, which has greatly influenced him as a systems thinker.

David Whiting is a data science director and head of training at H2O.ai. He has a PhD in statistics from Texas A&M University and over 25 years of professional experience in academia, consulting, and industry. He has built and led data science teams in financial services and other regulated enterprises.

About the reviewers

Jan Gamec is a lead software engineer at H2O.ai and one of the top contributors to a state-of-the-art AutoML platform called Driverless AI. In the past decade, he has contributed to various projects, focusing on machine learning, cryptography, and web technologies, either in the public or academic sector. Jan holds a master's degree in machine learning and computer science from CTU, Czech Republic, with the main focus of interest being genetic programming, neural networks, and reinforcement learning.

Jagadeesh Rajarajan has over 10 years of experience in building scalable data science systems. He has rich domain knowledge in the following areas: search relevance (information retrieval), recommender systems, AI for customer engagement (acquisition, activation, and retention), MLOps, and interpretable machine learning systems.

Eric Gudgeon has worked on many large complex systems, built nationwide networks, and helped customers deploy highly scalable low-latency solutions. He has a passion for technology and finding creative solutions to problems.

Ondrej Bilek is a lead software engineer at H2O.ai and has rich experience designing and implementing machine learning platforms for Hadoop and Kubernetes. He led the development of Enterprise Steam and is currently working on the H2O AI Cloud.

Table of Contents

3
Fundamental Workflow – Data to Deployable Model

Section 2 – Building State-of-the-Art Models on Large Data Volumes Using H2O

4
H2O Model Building at Scale – Capability Articulation

5

Advanced Model Building – Part I

6

Advanced Model Building – Part II

7

Understanding ML Models

8

Putting It All Together

Section 3 – Deploying Your Models to Production Environments

9
Production Scoring and the H2O MOJO

10
H2O Model Deployment Patterns

Section 4 – Enterprise Stakeholder Perspectives

11
The Administrator and Operations Views

12
The Enterprise Architect and Security Views

Section 5 – Broadening the View – Data to AI Applications with the H2O AI Cloud Platform

13

Introducing H2O AI Cloud

14

H2O at Scale in a Larger Platform Context

Appendix
Alternative Methods to Launch H2O Clusters

Index

Other Books You May Enjoy

Preface

At this point in time, **machine learning** (**ML**) requires little introduction: it is both pervasive and transformative to businesses, non-profits, and scientific organizations. ML is built on data. We are all aware of the exponential growth of data collected each year, and the growing diversity of sources that generate this data. This book is about leveraging these massive data volumes to do ML. We call this *machine learning at scale* and define it on three pillars: building high-quality models on large to massive datasets, deploying them for scoring in diverse enterprise environments, and navigating multiple stakeholder concerns along the way. Here, scale considers both data volume and enterprise context, model building, and model deployment. In this book, we will show you, in practical terms, how H2O overcomes the many challenges of performing ML at scale.

The book starts with a general overview of the challenges of performing ML at scale, and how the H2O framework overcomes these challenges while producing high-quality models and enterprise-grade deployments. From there, it transitions to advanced treatment of model-building techniques and model deployment patterns using H2O at Scale. We then look at its technological underpinnings from the perspective of multiple enterprise stakeholders who need to understand, deploy, and maintain this system, and show how this relates to data scientist activities and needs. We finish by showing how H2O at Scale can be implemented on its own or as part of the larger and richly featured H2O AI Cloud platform, where it takes on exciting new levels of ML possibilities and business value.

By the end of this book, you'll have the knowledge needed to build high-quality explainable ML models from massive datasets, deploy these models to a great diversity of enterprise systems, and assemble state-of-the-art ML solutions that achieve unique forms of business value.

Who this book is for

This book is written for data scientists, ML engineers, system administrators, enterprise architects, and curious technologists who want to build and deploy ML models at scale using H2O. Those already familiar with H2O will learn advanced model-building techniques and deployment patterns, as well as the details of how H2O works under the hood. Students with knowledge of ML but little or no work experience will gain an understanding of how it is performed in the world of large enterprises. Basic knowledge of ML is recommended, and an understanding of Python is needed to follow code examples.

What this book covers

Chapter 1, *Opportunities and Challenges*, sets the context of the book. We will first define ML at scale around three areas: building high-quality models on large to massive datasets, deploying them for scoring in diverse enterprise environments, and navigating multiple stakeholder concerns along the way. We will then recognize the vast business opportunities and execution challenges of ML in this context. In this light, you will be introduced to how H2O overcomes these challenges with its **H2O-3**, **Sparkling Water**, **Enterprise Steam**, and **MOJO** technologies that form the H2O at Scale framework.

Chapter 2, *Platform Components and Key Concepts*, overviews each H2O component by describing where it fits in the ML life cycle, what its key features are, and how it overcomes the challenges of ML at scale. We then distill several key concepts from this overview. The goal of this chapter is to provide you with a foundational knowledge of how H2O at Scale works before you learn how to implement it.

Chapter 3, *Fundamental Workflow – Data to Deployable Model*, shows the minimal steps needed to build and deploy models with the H2O at Scale framework. Think of this as a `Hello World` example, with each step explained. You have alternatives to implementing these steps, and they will be explored. At this point in the book, we will end our general overview and move on to advanced topics.

Chapter 4, *H2O Model Building at Scale – Capability Articulation*, starts our model-building focus and is of interest primarily to data scientists. In this chapter, we familiarize ourselves with H2O's extensive range of modeling capabilities, from data ingestion and manipulation to algorithms, model training, evaluation, and explainability techniques. Think of this chapter as the *what* of H2O model building, and the next chapters as an advanced treatment of the *how* and *why*.

Chapter 5, *Advanced Model Building – Part 1*, introduces you to the advanced model-building topics that a data scientist considers when building enterprise-grade models. We discuss data-splitting options, compare modeling algorithms, present a two-stage grid-search strategy for hyperparameter optimization, introduce H2O AutoML for automatically fitting multiple algorithms to data, and investigate feature engineering options for improving model performance. By the end of this chapter, you should be able to build an enterprise-scale, optimized, and predictive model using one or more supervised learning algorithms available within H2O.

Chapter 6, *Advanced Model Building – Part II*, continues our advanced model-building topics by showing how to build H2O supervised learning models within an Apache Spark pipeline, reviewing H2O's unsupervised learning methods, discussing best practices for updating H2O models, and introducing requirements to ensure H2O model reproducibility.

Chapter 7, *Understanding ML Models*, outlines a set of capabilities within H2O for explaining ML models. Building a model that predicts well is not enough. A critical step before putting any model into production is understanding how it makes decisions. We discuss selecting appropriate model metrics, using multiple diagnostics to build trust in a model, and using global and local explanations with model performance metrics to choose the best among a set of candidate models. This includes an evaluation of tradeoffs between model performance, speed of scoring, and assumptions met in a candidate model.

Chapter 8, *Putting It All Together*, starts the way most data science projects do: with raw data and a general business objective. We refine both the data and problem statement to be one that is relevant to the business and can be answered by the available data. We engineer a variety of features, creating and evaluating multiple candidate models until we arrive at a final model. We evaluate the final model and illustrate the preparation steps required for model deployment. The treatment in this chapter accurately reflects the job of a data scientist in the enterprise.

Chapter 9, *Production Scoring and the H2O MOJO*, starts our focus on model deployment. ML engineers, enterprise architects, software developers, and general technologists will be particularly interested in this chapter. You will become familiar with the strengths of H2O's MOJO as a scoring artifact, and how easily it can be deployed to a great diversity of enterprise systems. You will finish by writing a batch file scoring program that embeds a MOJO to demonstrate this flexibility.

Chapter 10, *H2O Model Deployment Patterns*, explores the many ways a MOJO can be deployed. You will first overview a diverse sampling of possible deployment patterns, and then drill down to implementation details of each. The patterns cover real-time streaming and batch scoring on a variety of specialized H2O scoring software, third-party integrations, and your own custom-built systems.

Chapter 11, The Administrator and Operations Views, starts our focus on enterprise stakeholder perspectives of ML at scale with H2O. Although focused on enterprise stakeholder activities and concerns, data scientists are shown how they relate to their own activities. In this chapter, system administrators and operators will learn in detail how Enterprise Steam is configured, and how users are secured and managed so data scientists can self-provision environments in a governed way. We will also identify operations activities around maintaining and troubleshooting H2O workloads and components.

Chapter 12, The Enterprise Architect and Security Views, covers the enterprise architect and security perspectives of H2O at Scale components. You will understand in detail the implementation alternatives of H2O and how the components integrate, communicate, and deploy. You will see that the H2O at Scale framework can be deployed on its own or as a member of the much larger **H2O AI Cloud**, which we cover in the next chapter.

Chapter 13, Introducing H2O AI Cloud, overviews H2O.ai's full end-to-end ML life cycle platform. The H2O at Scale framework and everything covered in the book to this point is a smaller subset of the H2O AI Cloud. In this chapter, we will overview H2O AI components and their key features, including four specialized **model-building engines**, a full-featured **MLOps** and **Feature Store**, and an open source low-code SDK to build and integrate **AI Apps** and host them on an **Appstore**.

Chapter 14, H2O at Scale in a Larger Platform Context, finishes the book by taking everything we have learned and showing how the H2O at Scale framework acquires categorically new and exciting possibilities when used as a part of the H2O AI Cloud. We provide examples of these possibilities and then present a reference enterprise integration framework using H2O for you to imagine your own possibilities.

Appendix, Alternative Methods to Launch H2O Clusters, shows different ways you can create H2O environments to run the code samples in this book.

To get the most out of this book

To run code samples, you will need to set up an H2O environment. The *Appendix* shows you the three ways to do this. Software versions are described there.

If you are using the digital version of this book, we advise you to type the code yourself or access the code from the book's GitHub repository (a link is available in the next section). Doing so will help you avoid any potential errors related to the copying and pasting of code.

From a conceptual standpoint, we introduce broad ML concepts before relating how H2O implements them. Therefore, you should be able to understand discussions in the book from your larger framework of ML understanding.

Download the example code files

You can download the example code files for this book from GitHub at https://
github.com/PacktPublishing/Machine-Learning-at-Scale-with-H2O.
If there's an update to the code, it will be updated in the GitHub repository.

We also have other code bundles from our rich catalog of books and videos available at
https://github.com/PacktPublishing/. Check them out!

Download the color images

We also provide a PDF file that has color images of the screenshots and diagrams used in
this book. You can download it here: https://packt.link/sDmtM.

Conventions used

There are a number of text conventions used throughout this book.

Code in text: Indicates code words in text, database table names, folder names,
filenames, file extensions, pathnames, dummy URLs, user input, and Twitter handles. Here
is an example: "The h2o.explain and h2o.explain_row methods bundle a set of
explainability functions and visualizations for global and local explanations, respectively."

A block of code is set as follows:

```
from pysparkling import *
import h2o
hc = H2Ocontext.getOrCreate()
hc
```

Any command-line input or output is written as follows:

```
PYSPARK_DRIVER_PYTHON="ipython" \
PYSPARK_DRIVER_PYTHON_OPTS="notebook" \
bin/pysparkling
```

Bold: Indicates a new term, an important word, or words that you see onscreen. For
instance, words in menus or dialog boxes appear in **bold**. Here is an example: "Under the
Admin menu in Flow, the top three options are **Jobs**, **Cluster Status**, and **Water Meter**."

> **Tips or Important Notes**
> Appear like this.

Get in touch

Feedback from our readers is always welcome.

General feedback: If you have questions about any aspect of this book, email us at customercare@packtpub.com and mention the book title in the subject of your message.

Errata: Although we have taken every care to ensure the accuracy of our content, mistakes do happen. If you have found a mistake in this book, we would be grateful if you would report this to us. Please visit www.packtpub.com/support/errata and fill in the form.

Piracy: If you come across any illegal copies of our works in any form on the internet, we would be grateful if you would provide us with the location address or website name. Please contact us at copyright@packt.com with a link to the material.

If you are interested in becoming an author: If there is a topic that you have expertise in and you are interested in either writing or contributing to a book, please visit authors.packtpub.com.

Share Your Thoughts

Once you've read *Machine Learning at Scale with H2O*, we'd love to hear your thoughts! Scan the QR code below to go straight to the Amazon review page for this book and share your feedback.

https://packt.link/r/1800566018

Your review is important to us and the tech community and will help us make sure we're delivering excellent quality content.

Section 1 – Introduction to the H2O Machine Learning Platform for Data at Scale

This section provides a general background of **machine learning** (ML) at scale with H2O. We will define ML at scale, focus on its challenges, and then see how H2O overcomes these challenges. We will then overview each H2O component to better understand its purpose and how it works from a technical standpoint. We will then put the components to work by implementing a minimal workflow. After this section, we will be ready to dive into advanced topics and techniques.

This section comprises the following chapters:

- *Chapter 1, Opportunities and Challenges*
- *Chapter 2, Platform Components and Key Concepts*
- *Chapter 3, Fundamental Workflow – Data to Deployable Model*

1
Opportunities and Challenges

Machine Learning (**ML**) and data science are winning a popularity contest of sorts, as witnessed by their headline coverage in the popular and professional press and by expanding job openings across the technology landscape. Students typically learn ML techniques using their own computers on relatively small datasets. Those who enter the field often find themselves in the much different setting of a large company buzzing with workers performing specialized job roles, while collaborating with others scattered across the nation or world. Both data science students and data science workers have a few key things in common – they are in an exciting and growing field that businesses deem ever more critical to their future, and the data they thrive on is becoming exponentially more abundant and diverse.

There are huge opportunities for ML in enterprises because the transformational impacts of ML on businesses, customers, patients, and so on are diverse, widespread, lucrative, and life-changing. A backdrop of urgency exists as well from competitors who are all attempting the same thing. Enterprises are thus incented to invest in significant ML transformations and to supply the necessary data, tooling, production systems, and people to journey toward ML success. But challenges loom large as well, and these challenges commonly revolve around scale. The challenges of scale take on many forms inherent to ML at an enterprise level.

In this chapter, we will define and explore the challenge of ML at scale by covering the following main topics:

- ML at scale

- The ML life cycle and three challenge areas for ML at scale

- H2O.ai's answer to these challenges

ML at scale

This book is about implementing ML at scale and how to use H2O.ai technology to succeed in doing so. What specifically do we mean by ML at scale? We can see three contexts and challenges of scale during the ML life cycle – building models from large datasets, deploying these models in enterprise production environments, and executing the full range of ML activities within the complexities of enterprise processes and stakeholders. This is summarized in the following figure:

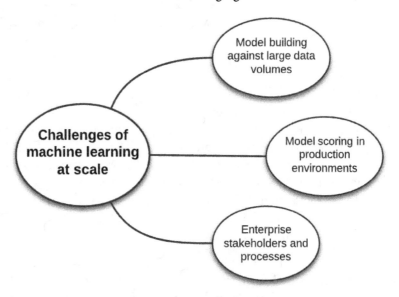

Figure 1.1 – The challenges of ML at scale

Let's drill down further on these challenges. Before doing so, we will oversee a generic conception of the ML life cycle, which will be useful as a reference throughout the book.

The ML life cycle and three challenge areas for ML at scale

The ML life cycle is a process that data scientists and enterprise stakeholders follow to build ML models and put them into production environments, where they make predictions and achieve value. In this section, we will define a simplified ML life cycle and elaborate on two broad areas that present special challenges for ML at scale.

A simplified ML life cycle

We will use the following ML life cycle representation. The goal is to achieve a simplified depiction that we can all recognize as central to ML while avoiding attempts at a canonical definition. Let's use it as our working framework for discussion:

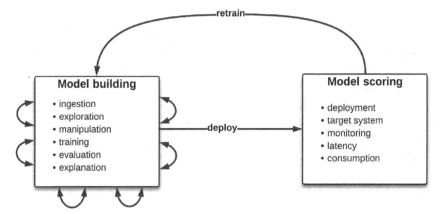

Figure 1.2 – A simplified ML life cycle

The following is a brief articulation.

Model building

Model building is a highly iterative process with frequent and unpredictable feedback loops along the way toward building a predictive model that is worthy of deploying in a business context. The steps can be summarized as follows:

- **Data ingestion**: Data is pulled from sources or a storage layer in the model building environment. There is often significant work onward from here in finding and accessing potentially useful data sources and transforming the data into a useable form. Typically, this is done as part of a larger data pipeline and architecture.

- **Data exploration**: Data is explored to understand its qualities (for example, data profiling, correlation analysis, outlier detection, and data visualization).

- **Data manipulation**: Data is cleaned (for example, the imputation of missing data, the reduction of categorical features, and normalization) and new features are engineered.

- **Model training**: An ML algorithm, scoring metric, and validation method are selected, and the model is tuned across a range of hyperparameters and tested against a test dataset.

- **Model evaluation and explainability**: A fit of the model is diagnosed for performance metrics, overfitting, and other diagnostics; model explainability is used to validate against domain knowledge, to explain the model decisions at individual and global levels, and to guard against institutional risks such as unfair bias against demographic groups.

- **Model deployment**: The model is deployed as a scoring artifact to a software system and live scoring is made.

- **Model monitoring**: The model is monitored to detect whether the data fed into it changes over time compared to the distribution of data it was trained on. This is called data drift and usually leads to the decreased predictive power of the model. This usually triggers the need to retrain the model with a more current dataset and then redeploy the updated model. The model may also be monitored for other patterns, such as whether it is biasing decisions against a particular demographic group and whether malicious attacks are being made to try to cause the model to malfunction.

As mentioned, a key property in the workflow is the unknown number and sequence of iteration pathways taken between these steps before a model is deployed or before the project is deemed unsuccessful in reaching that stage.

The model building challenge – state-of-the-art models at scale

Let's, for now, define a large dataset as any dataset that exceeds your ability to build ML models on your laptop or local workstation. It may be too large because your libraries simply crash or because they take an unreasonable amount of time to complete. This may occur during model training or during data ingestion, exploration, and manipulation.

We can see four separate challenges of building ML models from large data volumes, with each contributing to a larger problem in general that we call the *friction of iteration*. This is represented in the following diagram:

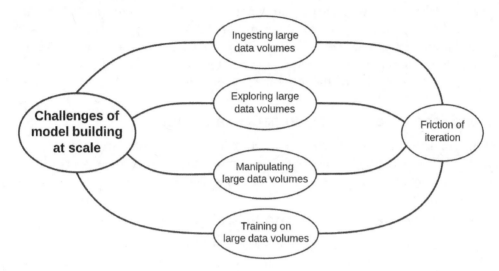

Figure 1.3 – The challenge of model building with large data volumes

Let's elaborate on this.

Challenge one – data size and location

Enterprises collect and store vast amounts of diverse data and that is a boon to the data scientist looking to build accurate models. These datasets are either stored across many systems or centralized in a common storage layer (data lake) such as the **Hadoop Distributed File System** (**HDFS**) or **AWS S3**. Architecting and making data available to internal consumers is a major effort and challenge for an enterprise. However, the data scientist starting the ML life cycle with large datasets typically cannot move that data, once it becomes accessible, to a local environment due to either security reasons or high volume of data.. The consequence is that the data scientist must either do one of the following:

- Move operations on the data (in other words, move the compute) to the data itself.

- Move data to a high-compute environment that they are authorized to use.

Challenge two – data size and data manipulation

Manipulating data can be compute-intensive, and attempting to do so against insufficient resources either will cause the compute to fail (for example, the script, library, or tool will crash) or take an unreasonably long amount of time. Who wants to wait 10 hours to join and filter table data when it can be done in 10 minutes? What you might consider an unreasonable amount of time is obviously relative to the dataset size; terabytes of data will always take longer to process than a few megabytes. Regardless, the speed of your data processing is critical to reducing the sum time of your iterations.

Challenge three – data size and data exploration

Challenges of data size during data exploration are identical to those during data manipulation. The data may be so large that your processing crashes or takes an unreasonable amount of time to complete while exploring models.

Challenge four – data size and model training

ML algorithms are extremely *compute-intensive* because they step through each record of a dataset and perform complex calculations each time, and then iterate these calculations against the dataset repeatedly to optimize toward a training metric and thus learn a predictive mathematical pattern among the noise. Our compute environment is particularly pressured during model training.

Up until now, we have been discussing dataset size in relative terms; that is, large data volumes are those that cause operations on them to either fail or take a long time to complete in a given compute environment.

In absolute terms, data scientists often explore the largest dataset possible to understand it and then **sample** it for model training. Others always try to use the largest dataset for model training. However, accurate models can be built from 10 GB or less of sampled or unsampled data.

The key to proper use of sampling is that you have followed appropriate statistical and theoretical practices, and not that you are forced to do so because your ML processing will crash or take a long time to complete due to large data volumes. The latter is a bad practice that produces inferior models and H2O.ai overcomes this by allowing model building with massive data volumes.

There are also cases when data sampling may not lead to an acceptable model. In other words, the data scientist may need hundreds of gigabytes or a terabyte or more of data to build a valuable model. These are cases when the following applies:

- The data scientist does not trust the sampling to produce the best model and feels that each small gain in lift warrants the use of the full dataset.

- The data scientist does not want to segment the data into separate datasets and thus separate model building exercises, or the larger stakeholder group wants a single model in production that predicts against all segments versus many that each predicts against a single segment.

- The data is highly dimensional, sparse, or both. In this case, a large number of records are needed to reduce variance and overfitting to a training dataset. This type of dataset is typical for anomaly detection, recommendation engines, predictive maintenance, security threat detection, personalized medicine, and so on. It is worth noting that the future will bring us more and more data, and thus highly dimensional and sparse datasets will become more common.

- The data is extremely imbalanced. The target variable is very rare in the dataset and a massive dataset is needed to avoid underfitting, overfitting, or weighting the target variable from these infrequent records.

- The data is highly volatile. Each subset of data that is collected is unrepresentative of the others and thus sampling or cross-validation folds may not be representative. Time series forecasting may be particularly sensitive to this problem, especially when forecast categories are highly granular (for example, yearly, monthly, daily, and hourly) against a single validation dataset.

The friction of iteration

Model building is a highly iterative process and anything that slows it down we call the friction of iteration. These causes can be due to the challenges of working with large data volumes, as previously discussed. They can also arise from simple workflow patterns such as switching among systems between each iteration or launching new environments to work on an iteration.

Any slowness during a single iteration may seem acceptable but when multiplied across the seemingly endless iterations from the project beginning to failure or success, the cost in time from this friction becomes significant, and reducing friction can be valuable. As we will see in the next section, slow model building delays the main goal of ML in an enterprise – achieving business value.

The business challenge – getting your models into enterprise production systems

The bare truth about ML initiatives is that they do not really achieve value until they are deployed to a live scoring environment. Models must meet evaluation criteria and be put into production to be deemed successful. Until that happens, from a business standpoint, little is achieved. This may seem a bit harsh, but it is typically how success is defined in data science initiatives. The following diagram maps this thinking onto the ML life cycle:

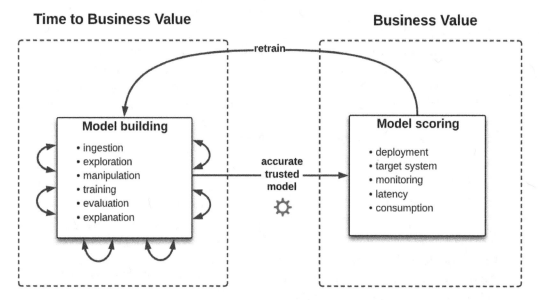

Figure 1.4 – The ML life cycle value chain

The friction of iteration from this view is thus a cost. Time taken to iterate through model building is time taken from getting business results. In other words, lower friction translates to less time to build and deploy a model to achieve business value, and more time to work on other problems and thus more models per quarter or year.

From the same point of view, **time to deploy a model** is viewed as a cost for similar reasons. The model deployment step may seem like a simple one-step sequence of transitioning the model to DevOps, but typically it is not. Anything that makes a model easier and more repeatable to deploy, document, and govern helps businesses achieve value sooner.

Let's now continue expanding on a larger landscape of enterprise stakeholders that data scientists must work with to build models that ultimately achieve business value.

The navigation challenge – navigating the enterprise stakeholder landscape

The data scientist in any enterprise does not work in isolation. There are multiple stakeholders who become involved directly in the ML life cycle or, more broadly, in the business cycle of initiating and consuming ML projects. Who might some of these stakeholders be? At a bare minimum, they include the business stakeholder who funded the ML project, the administrator providing the data scientist with permissions and capabilities, the DevOps or engineering team members who are responsible for model deployment and the infrastructure supporting it, perhaps marketing or sales associates whose functions are impacted directly by the model, and any other representatives of the internal or external consumers of the model. In more heavily regulated industries such as banking, insurance, or pharmaceuticals, these might include representatives or offices of various audit and risk functions – data risk, code risk, model risk, legal risk, reputational risk, compliance, external regulators, and so on. The following figure shows a general view:

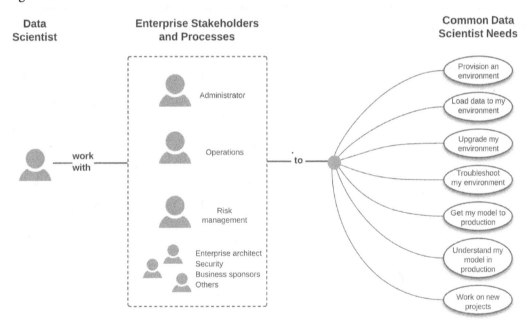

Figure 1.5 – Data scientists working with enterprise stakeholders and processes

Stakeholder interaction is thus complex. What leads to this complexity? Obviously, the specialization and siloing of job functions make things complex, and this is further amplified by the scale of the enterprise. A larger dynamic of creating repeatable processes and minimizing risk contributes as well. Explaining this complexity is the task of a different book, but its reality in the enterprise is inescapable. To a data scientist, the ability to recognize, influence, negotiate with, deliver to, and ultimately build trust with these various stakeholders is imperative to successful ML solutions at scale.

Now that we have understood the ML life cycle and the challenges inherent in its successful execution at scale, it is time for a brief introduction to how H2O.ai solves these challenges.

H2O.ai's answer to these challenges

H2O.ai provides software to build ML models at scale and overcome the challenges of doing so – model building at scale, model deployment at scale, and dealing with enterprise stakeholders' concerns and inherent friction along the way. These components are described in brief in the following diagram:

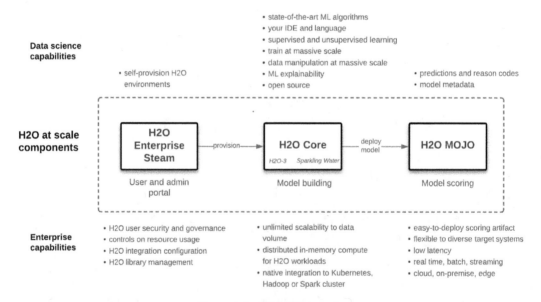

Figure 1.6 – H2O ML at scale

Subsequent chapters of this book elaborate on how these components are used to build and deploy state-of-the-art models within the complexities of the enterprise environment.

Let's try to understand these components at first glance:

- **H2O Core**: This is open source software that distributes state-of-the-art ML algorithms and data manipulations over a specified number of servers on Kubernetes, Hadoop, or Spark environments. Data is partitioned in memory across the designated number of servers and ML algorithm computation is run in parallel using it.

  This architecture creates horizontal scalability of model building to hundreds of gigabytes or terabytes of data and generally fast processing times at lower data volumes. Data scientists work with familiar IDEs, languages, and algorithms and are abstracted away from the underlying architecture. Thus, for example, a data scientist can run an XGBoost model in Python from a Jupyter notebook against 500 GB of data in Hadoop, similar to doing so with data loaded into their laptop.

  H2O Core is often referred to as **H2O Open Source** and comes in two forms, **H2O-3** and **Sparkling Water**, which we will elaborate on in subsequent chapters. H2O Core can be run as a scaled-down sandbox on a single server or laptop.

- **H2O Enterprise Steam**: This is a web UI or API for data scientists to self-provision and manage their individual H2O Core environments. Self-provisioning includes auto-calculation of horizontal scaling based on user inputs that describe the data. Enterprise Steam is also used by administrators to manage users, including defining boundaries for their resource consumption, and to configure H2O Core integration against Hadoop, Spark, or Kubernetes.

- **H2O MOJO**: This is an easy-to-deploy scoring artifact exportable from models built from H2O Core. MOJOs are low latency (typically < 100 ms or faster) Java binaries that can run on any **Java Virtual Machine (JVM)** and thus serve predictions on diverse software systems, such as REST servers, database clients, Amazon SageMaker, Kafka queues, Spark pipelines, Hive **user-defined functions (UDFs)**, and **Internet of Things (IoT)** devices.

- **APIs**: Each component has a rich set of APIs so that you can automate workflows, including **continuous integration and continuous delivery (CI/CD)** and retraining pipelines.

The focus of this book is on building and deploying state-of-the-art models at scale using H2O Core with help from Enterprise Steam and deploying those models as MOJOs within the complexities of enterprise environments.

> **H2O at Scale and H2O AI Cloud**
>
> We refer to **H2O at scale** in this book as H2O Enterprise Steam, H2O Core, and H2O Mojo because it addresses the ML at scale challenges described earlier in this chapter, especially through the distributed ML scalability that H2O Core provides for model building.
>
> Note that H2O.ai offers a larger end-to-end ML platform called the **H2O AI Cloud**. The H2O AI Cloud integrates a hyper-advanced AutoML tool (called **H2O Driverless AI**) and other model building engines, an MLOps scoring, monitoring, and governance environment (called **H2O MLOps**), and a low-code software development kit, or SDK (called **H2O Wave**) with H2O API hooks to build AI applications that publish to the **App Store**. It also integrates H2O at scale as defined in this book.
>
> H2O at scale can be deployed as standalone or as part of the H2O AI Cloud. As a standalone implementation, Enterprise Steam is not in fact required, but for reasons elaborated on later in this book, Enterprise Steam is deemed essential for enterprise implementations.
>
> The majority of this book is focused on H2O at scale. The last part of the book will extend our understanding to the H2O AI Cloud and how H2O at scale components can leverage this larger integrated platform and vice versa.

Summary

In this chapter, we have set the stage for understanding and implementing ML at scale using H2O.ai technology. We have defined multiple forms of scale in an enterprise setting and articulated the challenges to ML from model building, model deployment, and enterprise stakeholder perspectives. We have anchored these challenges ultimately to the end goal of ML – providing business value. Finally, we briefly introduced H2O at scale components used by enterprises to overcome these challenges and achieve business value.

In the next chapter, we'll start to understand these components in greater technical detail so that we can start writing code and doing data science.

2
Platform Components and Key Concepts

In this chapter, we will gain a fundamental understanding of the components of H2O's machine learning at scale technology. We will view a simple code example of H2O machine learning, understand what it does, and identify any problems the example has with machine learning at an enterprise scale. This *Hello World* code example will serve as a simple representation in which to build our understanding further.

We will overview each H2O component of machine learning at scale, identify how each component achieves scale, and identify how each component relates to our simple code snippet. Then, we will tie these components together into a reference machine learning workflow using these components. Finally, we will focus on the underlying key concepts that arise from these components. The understanding obtained in this chapter will be foundational to the rest of the book, where we will be implementing H2O technology to build and deploy state-of-the-art machine learning models at scale in an enterprise setting.

In this chapter, we're going to cover the following main topics:

- Hello World – the H2O machine learning code
- The components of H2O machine learning at scale
- The machine learning workflow using these H2O components
- H2O key concepts

Technical requirements

For this chapter, you will need to install H2O-3 locally to run through a bare minimum *Hello World* workflow. To implement it, follow the instructions in the *Appendix*. Note that we will use the Python API throughout the book, so follow the instructions to install it in Python.

Hello World – the H2O machine learning code

H2O Core is designed for machine learning at scale; however, it can also be used on small datasets on a user's laptop. In the following section, we will use a minimal code example of H2O-3 to build a machine learning model and export it as a deployable artifact. We will use this example to serve as the most basic unit to understand H2O machine learning code, much like viewing a human stick figure to begin learning about human biology.

Code example

Take a look at the code examples that follow. Here, we are writing in Python, which could be from Jupyter, PyCharm, or another Python client. We will learn that R and Java/Scala are alternative languages in which to write H2O code.

Let's start by importing the H2O library:

```
import h2o
```

Recall from the documentation that this has been downloaded from H2O and installed in the client or an IDE environment. This h2o package allows us to run H2O in-memory distributed machine learning from the IDE using the H2O API written in Python.

Next, we create an H2O cluster:

```
h2o.init(ip="localhost", port=54323)
```

The preceding line of code creates what is called An **H2O cluster**. This is a key concept underlying H2O's model building technology. It is a distributed in-memory architecture. In the *Hello World* case, the H2O cluster will be created on the laptop as localhost and will not be distributed. We will learn more about the H2O cluster in the *H2O key concepts* section of this chapter.

The `ip` and `port` configurations that are used to start the H2O cluster should provide sufficient clues that the H2O code will be sent via an API to the compute environment, which could be inside a data center or the cloud for an enterprise environment. However, here, it is on our localhost.

Then, we import a dataset:

```
loans = h2o.import_file("https://raw.githubusercontent.com/
PacktPublishing/Machine-Learning-at-Scale-with-H2O/main/chapt2/
loans-lite.csv")
```

Now we explore the dataset:

```
loans.describe()
```

This is a minimal amount of data exploration. It simply returns the number of rows and columns.

Okay, now let's prepare the data for our model:

```
train, validation = loans.split_frame(ratios=[0.75])
label = "bad_loan"
predictors = loans.col_names
predictors.remove(label)
```

We have split the data into training and validation sets, with a 0.75 proportion for training. We are going to predict whether a loan will be bad or not (that is, whether it will default or not) and have identified this column as the label. Finally, we define the columns used to predict bad loans by using all columns in the dataset except the bad loan column.

Now, we build the model:

```
from h2o.estimators import H2OXGBoostEstimator
param = {"ntrees" : 25, "nfolds" : 10}
xgboost_model = H2OXGBoostEstimator(**param)
xgboost_model.train(x = predictors,
                    y = label,
                    training_frame = train,
                    validation_frame = validation)
```

We have imported H2O's **XGBoost** module and configured two hyperparameters for it. Then, we started the model training by inputting references into the predictor column, label column, training data, and testing data.

XGBoost is one of many widely recognized and extensively used machine learning algorithms packaged in the h2o module. The H2O API exposed by this module will run the XGBoost model in H2O's architecture on the enterprise infrastructure, as we will learn later. Regarding hyperparameters, we will discover that H2O offers an extensive set of hyperparameters to configure for each model.

When the model finishes, we can export the model using one line of code:

```
xgboost_model.download_mojo(path="~/loans-model", get_genmodel_
jar=True)
```

The exported scoring artifact is now ready to pass to DevOps to deploy. The get_ genmodel_jar=True parameter triggers the download to include h2o-genmodel. jar. This is a library used by the model for scoring outside of an H2O cluster, that is, in a production environment. We will learn more about productionizing H2O models in *Section 3 – Deploying Your Models to Production Environments*.

We are done with model building, for now. So, we will shut down the cluster:

```
h2o.cluster().shutdown()
```

This frees up the resources that the H2O cluster has been using.

Bear in mind that this is a simple *Hello World* H2O model building example. It is meant to do both of the following:

- Give a bare minimum introduction to H2O model building.

- Serve as a basis to discuss issues of scale in the enterprise, which we will do in the next section.

In *Section 2 – Building State-of-the-Art Models on Large Data Volumes Using H2O*, we will explore extensive techniques to build highly predictive and explainable models at scale. Let's start our journey by discussing some issues of scale that our *Hello World* example exposes.

Some issues of scale

This *Hello World* code will not scale well in an enterprise setting. Let's revisit the code to better understand these scaling constraints.

We import the library in our IDE code:

```
import h2o
```

Most enterprises want to have some control over the versions of libraries that are used. Additionally, they usually want to provide a central platform to host and authenticate all users of a piece of technology and to have administrators manage that platform. We will discover that Enterprise Steam plays a key role in centrally managing users and H2O environments.

We initialize the H2O cluster:

```
h2o.init(ip="localhost", port=54323)
```

Machine learning at scale requires the distribution of compute resources across a server cluster to achieve horizontal scaling (that is, divide-and-conquer compute resources across many servers). Therefore, the IP address and port should point to a member of a server cluster and not to a single computer, as demonstrated in this example. We will see that H2O Core creates its own self-organized cluster that distributes and horizontally scales model building.

Since scaling is on the enterprise server cluster, which, typically, is used by many individuals and groups, enterprises want to control user access to this environment along with the number of resources consumed by users. But then what would prevent a user from launching multiple H2O clusters, using as many resources as possible on each, and thus, blocking resource availability from other users? Enterprise Steam manages H2O user and H2O resource consumption on the enterprise server cluster.

We import the dataset:

```
loans = h2o.import_file("https://raw.githubusercontent.com/
PacktPublishing/Machine-Learning-at-Scale-with-H2O/main/chapt2/
loans-lite.csv")
```

Data at large volumes takes an exceedingly long time to move over the network, taking hours or days to complete a transfer, or it could time out beforehand. Computation during model building at scale should occur where the data resides to prevent this bottleneck in data movement. We will discover that H2O clusters that are launched on the enterprise system ingest data from the storage layer directly into server memory. Because data is partitioned across the servers that comprise an H2O cluster, data ingest occurs in parallel to those partitions.

We will see how Enterprise Steam centralizes user authentication and how the user's identity is passed to the enterprise system where its native authorization mechanism is honored.

We train the model:

```
xgboost_model.train(x = predictors,
                    y = label,
                    training_frame = train,
                    validation_frame = validation)
```

Of course, this is the heart of the model building process and, likewise, the focus of much of this book: how to build world-class machine learning models against large data volumes using H2O's extensive machine learning algorithm and model building capabilities.

We download the deployable model:

```
xgboost_model.download_mojo(path="~/loans-model", get_genmodel_
jar=True)
```

Bear in mind that, from a business standpoint, value is not achieved until a model is exported and deployed into production. Doing so involves the complexities of multiple enterprise stakeholders. We will learn how the design and capabilities of the exported **MOJO (Model Object, Optimized)** facilitate the ease of deployment to diverse software systems involving these stakeholders.

We shut down the H2O cluster:

```
h2o.cluster().shutdown()
```

An H2O cluster uses resources and should be shut down when not in use. If this is not done, other users or jobs on the enterprise system could be competing for these resources and, consequently, become impacted. Additionally, fewer new users can be added to the system before the infrastructure must be expanded. We will see that Enterprise Steam governs how H2O users consume resources on the enterprise system. The resulting gain in resource efficiency allows H2O users and their work to scale more effectively on a given allocation of infrastructure.

Now that we have run our *Hello World* example and explored some of its issues regarding scale, let's move on to gain an understanding of H2O components for machine learning model building and deployment at scale.

The components of H2O machine learning at scale

As introduced in the previous chapter and emphasized throughout this book, H2O machine learning overcomes problems of scale. The following is a brief introduction of each component of H2O machine learning at scale and how each overcomes these challenges.

H2O Core – in-memory distributed model building

H2O Core allows a data scientist to write code to build models using well-known machine learning algorithms. The coding experience is through an H2O API expressed in Python, R, or Java/Scala language and written in their favorite client or IDE, for example Python in a Jupyter notebook. The actual computation of model building, however, takes place on an enterprise server cluster (not the IDE environment) and leverages the server cluster's vast pool of memory and CPUs needed to run machine learning algorithms against massive data volumes.

So, how does this work? First, data used for model building is partitioned and distributed in memory by H2O on the server cluster. The IDE sends H2O instructions to the server cluster. A server in the cluster receives these instructions and distributes them to the other servers in the cluster. The instructions are run in parallel on the partitioned in-memory data. The server that received the instructions gathers and combines the results and sends them back to the IDE. This is done repeatedly as code is sequenced through the IDE.

This *divide and conquer* approach is fundamental to H2O model building at scale. A unit of H2O divide and conquer architecture is called an H2O cluster and is elaborated as a *key concept* later in the chapter. The result is rapid model building on large volumes of data.

The key features of H2O Core

Some of the key features of H2O Core are as follows:

- **Horizontal scaling**: Data operations and machine learning algorithms are distributed in parallel and in memory, with additional optimizations such as a distributed key/value store to rapidly access data and objects during model building.

- **Familiar experience**: Data scientists use familiar languages and IDEs to write H2O API code, as we have just done.

- **Open source**: H2O Core is open source.

- **Wide range of file formats**: H2O supports a wide range of source data formats.

- **Data manipulation**: The H2O API includes a wide range of tasks commonly performed to prepare data for machine learning. Sparkling Water (covered in the next section) extends data engineering techniques to Spark.

- **Well-recognized machine learning algorithms**: H2O uses a wide range of well-recognized supervised and unsupervised machine learning algorithms.

- **Training, testing, and evaluation**: Extensive techniques in cross-validation, grid search, variable importance, and performance metrics are used to train, test, and evaluate models; this also includes model checkpointing capabilities.

- **Automatic Machine Learning (AutoML)**: The H2O Core AutoML API provides a simple wrapper function to concisely automate the training and tuning of multiple models, including stacked ensembling, and present results in a leaderboard.

- **Model explainability**: It offers extensive local and global explainability methods and visualizations for single models or those involved in AutoML, all from a single wrapper function.

- **AutoDoc**: It enables the automated generation of standardized Word documents, extensively describing model building and explainability in detail; note that AutoDoc is not available as a free open source platform.

- **Exportable scoring artifact (MOJO)**: It uses a single line of code to export the model as a deployable scoring artifact (model deployment will be discussed in greater detail in *Section 3 – Deploying Your Models to Production Environments*).

- **H2O Flow Web UI**: This is an optional web-based interactive UI to guide users through the model building workflow in an easy yet rich point-and-click experience, which is useful for the rapid experimentation and prototyping of H2O models.

H2O-3 and H2O Sparkling Water

H2O Core comes in two flavors: **H2O-3** and **H2O Sparkling Water**.

H2O-3 is H2O Core, as described in the previous section. H2O Sparkling Water is H2O-3 wrapped by Spark integration. It is identical to H2O-3 along with the following additional capabilities:

- **Seamless integration of Spark and H2O API code**: The user writes both Spark and H2O code in the same IDE; for example, using SparkSQL code to engineer data and H2O code to build world-class models.

- **Conversion between H2O and Spark DataFrames**: H2O and Spark DataFrames interconvert as part of the seamless integration; therefore, the results of SparkSQL data munging can be used as input to H2O model building.

- **Spark engine**: Sparkling Water runs as a native Spark application on the Spark framework.

H2O-3 and Sparkling Water are the model building alternatives of the more general H2O Core. The concept of the H2O cluster launched on the larger enterprise server cluster is similar for both H2O Core flavors, though some implementation details differ, which are essentially invisible to the data scientist. As mentioned, Sparkling Water is particularly useful for integrating Spark data engineering and H2O model building workflows.

H2O Enterprise Steam – a managed, self-provisioning portal

Enterprise Steam provides a centralized web UI and API for data scientists to initialize and terminate their H2O environments (called H2O clusters) and for administrators to manage H2O users and H2O integration with the enterprise server cluster.

The key features of Enterprise Steam

The key features of Enterprise steam are as follows:

- **Data science self-provisioning**: This is an easy, UI-based way for data scientists to manage their H2O environments.

- **Central access point for all H2O users**: This creates the ease of H2O user management and a single entry point for H2O access to the enterprise server cluster.

- **Govern user resource consumption**: Administrators build profiles of resource usage boundaries that are assigned to users or user groups. This places limits on the number of resources a user can allocate on the enterprise server cluster.

- **Seamless security**: User authentication to Enterprise Steam flows through to the authorization of resources on the enterprise server cluster. Enterprise Steam authenticates against the same identity provider (for example, LDAP) that is used by the enterprise server cluster.

- **Configure integration**: The administrator configures the integration of H2O with the enterprise server cluster and identity provider.

- **Manage H2O Core versions**: The administrator manages one or more H2O Core versions that data scientists use to create H2O clusters for model building.

The H2O MOJO – a flexible, low-latency scoring artifact

The models built from H2O Core are exported as deployable scoring artifacts called H2O MOJOs. MOJOs can run in any JVM environment (except, perhaps, the very smallest edge devices).

In *Section 3 – Deploying Your Models to Production Environments*, we will learn that MOJOs are ready to deploy directly to H2O software as well as many third-party scoring solutions with no coding required. However, if you wish to directly embed MOJOs into your own software, there is a MOJO Java API to build Java helper classes to expose MOJO capabilities (for example, output reason codes in addition to a prediction) and to provide flexible integration with your scoring input and output.

MOJOs, out of all models, regardless of the machine learning algorithm used to build the model, are identical in construct. Therefore, deployment from a DevOps perspective is repeatable and automatable.

The key features of MOJOs

The key features of the MOJO are as follows:

- **Low latency**: Typically, this is less than 100 milliseconds for each scoring.
- **Flexible data speeds**: Mojos can make predictions on batch, real time, and streaming data (for example on entire database tables, as REST endpoints and Kafka topics, respectively, to name a few examples).

- **Flexible target systems**: This fits into JVM runtimes, including JDBC clients, **REST servers**, **AWS Lambda**, **AWS SageMaker**, **Kafka queues**, **Flink streams**, Spark pipelines including streaming, Hive UDF, Snowflake's external functions, and more. Target systems can be specialized H2O scoring software, third-party scoring software, or your own software. A common pattern is to deploy the MOJO to a REST server and consume its predictions via REST calls from a client application (for example, an Excel spreadsheet).

- **Explainability features**: In addition to predictions, you can receive K-Lime or Shapley reason codes from the MOJO during live scoring, and you can load the MOJO into H2O Core to score and inspect MOJO attributes.

- **Repeatable deployments**: MOJOs are easy to integrate into existing deployment automation (CI/CD) pipelines used by the organization for software deployment.

Note that there is an alternative to the H2O MOJO, called **POJO**, which is used for infrequent edge cases. This will be explored further in *Chapter 8*, *Putting It All Together*.

The workflow using H2O components

Now that we understand the roles and key features of H2O's machine learning at scale components, let's tie them together into a high-level workflow, as represented in the following diagram:

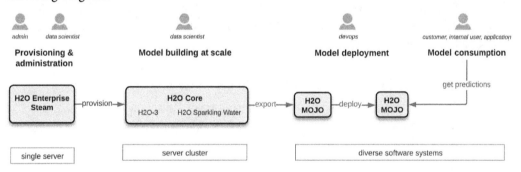

Figure 2.1 – A high-level machine learning at scale workflow with H2O

The workflow occurs in the following sequence:

1. The administrator configures **H2O Enterprise Steam**.

2. The data scientist logs into **H2O Enterprise Steam** and launches the **H2O Core** cluster (choosing either **H2O-3** or **H2O Sparkling Water**).

3. The data scientist uses their favorite client to build models using the Python, R, or Java/Scala language flavor of the H2O model building API. The data scientist uses a UI or IDE to authenticate to **H2O Enterprise Steam** and connect to the **H2O cluster** that was started on H2O Enterprise Steam.

4. The data scientist uses the IDE to iterate through the model building steps with H2O.

5. After the data scientist decides on the model to be deployed, **H2O AutoDoc** is generated, and **H2O MOJO** is exported from the IDE.

6. The data scientist either terminates the **H2O cluster** or waits for **H2O Enterprise Steam** to do so after the idle or absolute uptime duration has been exceeded. These durations have been configured in a resource profile assigned to the user by the administrator. Note that the terminated cluster checkpoints do work, and a new **H2O cluster** can always be launched to continue working from the termination point.

7. The model is exported as **H2O MOJO** and is deployed to any of a diverse set of hosting targets. The model is consumed in a business context and achievement of the business value begins.

H2O key concepts

In the following sections, we will identify and describe the key concepts of H2O that underlie the workflow steps of the previous section. These concepts are necessary to understand the rest of the book.

The data scientist's experience

The data scientist has a familiar experience in building H2O models at scale while being abstracted from the complexities of the infrastructure and architecture on the enterprise server cluster. This is further detailed in the following diagram:

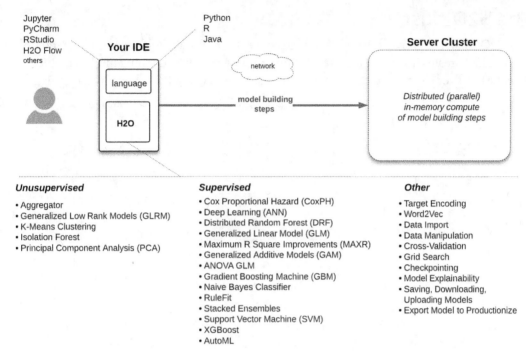

Figure 2.2 – Details of the data scientist's experience with H2O Core

Data scientists use well-known unsupervised and supervised machine learning techniques that scale across the enterprise's distributed infrastructure and architecture. These techniques are written with the H2O model building API, which is written in familiar languages (such as Python, R, or Java) using familiar IDEs (for example, Jupyter or RStudio).

> **H2O Flow – A Convenient, Optional UI**
>
> H2O generates its own web UI called H2O Flow, which is optional to use during model building. H2O Flow's UI focus and richness of features can be used for a full model building workflow or to leverage for handy tricks, as we will demonstrate in *Chapter 5, Advanced Model Building – Part 1.*

Therefore, the data scientist works in a familiar world that connects to a complex architecture to scale model building to large or massive datasets. We will explore this architecture in the next section.

The H2O cluster

The H2O cluster is perhaps the most central concept for all stakeholders to understand. It is how H2O creates its unit of architecture for building machine learning models on the enterprise server cluster. We can understand this concept using the following diagram:

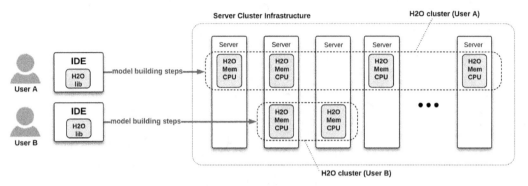

Figure 2.3 – The architecture of the H2O cluster

When a data scientist launches an H2O cluster, they specify the number of servers to distribute the work across (which is also known as the number of *nodes*), along with the amount of memory and CPUs to use for each node. We will learn that this can be done by configuring manually or by allowing Enterprise Steam to auto compute these specifications based on the volume of training data.

When the H2O cluster is launched, the IDE pushes H2O software (a single JAR file) to each specified number of nodes in the enterprise server cluster, where each node allocates the specified memory and CPU. Then, the H2O software organizes into a self-communicating cluster with one node elected as the leader that communicates with the IDE and coordinates with the remainder of the H2O cluster.

The data scientist connects to the launched H2O cluster from the IDE. Then, the data scientist writes the model building code. Each part of the code is translated by the H2O library in the IDE into instructions to the H2O cluster. Each instruction is sent, in sequence, to the leader node on the H2O cluster, which distributes it to other H2O cluster members where the instructions are executed in parallel. The leader node gathers and combines the results and sends them back to the IDE.

Here are some important notes to bear in mind:

- Data is ingested directly from the data source to the memory of the H2O nodes. Source data is partitioned between the H2O nodes and not duplicated among them. Data ingested from the storage layer (for example, S3, HDFS, and more) is done in parallel and, therefore, is fast. Data from external sources (for example, the GitHub repository and the JDBC database tables) is not done in parallel. In all cases, data does not pass through the IDE or the client.

- Each H2O cluster is independent and isolated from the others, including the data ingested into them. Thus, two users launching a cluster and using the same data source do not share data.

- We will see that administrators of Enterprise Steam assign upper limits on the number of concurrent clusters that users can launch, along with the amount of memory, CPU, and other resources a user can specify when launching a cluster.

- H2O clusters are static. Once launched, the number of nodes and the number of resources per node do not change until they are terminated, in which case the H2O cluster is torn down. If one of the nodes goes down, the H2O cluster must be restarted and model building steps from the IDE started from the beginning. For longer durations of work, H2O's checkpointing feature helps you to continue from a restore point.

Let's look at the life cycle of an H2O cluster, as shown in the following diagram:

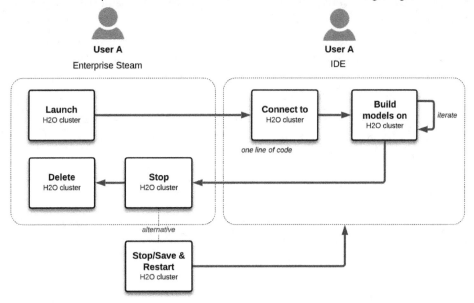

Figure 2.4 – The life cycle of the H2O cluster

Let's look at each of the stages of the life cycle, one by one, to understand how they work:

1. **Launch**: The data scientist launches an H2O cluster from the Enterprise Steam UI or API. H2O-3 or Sparkling Water is chosen. The H2O cluster size and resources (that is, the number of nodes, memory per node, and other configurations) are manually input, or they are automatically generated by Enterprise Steam based on the data volume input by the user. The H2O cluster is formed as described earlier.

2. **Connect to**: The data scientist switches to their IDE and connects to the H2O cluster by specifying its name.

3. **Build models on**: The data scientist builds models using H2O. The H2O library used in the IDE translates the H2O API code for each model building iteration into instructions. These are sent to the leader node and distributed across the H2O cluster.

4. **Stop**: The H2O cluster is shut down. Resources are released, and the H2O software is removed from each node of the H2O cluster. This can be done by the user from the IDE or can occur automatically after a duration of idle time or when the absolute running time of the H2O cluster has been exceeded (these durations were specified in the H2O cluster launch during step 1 of the life cycle). Though not running, information regarding this cluster is still available to the user (for example, the name, the H2O version, and the size).

    Stop/Save Data & Restart: This is an alternative to **Stop** and is possible when the Enterprise Steam administrator configures this option for a user or user group. In this case, when the H2O cluster is stopped, it saves data from the model building steps (that is, it saves the model building state) to the storage layer. When the cluster is restarted (using the same name as when it was launched), the cluster is launched and returned to its previous state.

5. **Delete**: This stops the cluster (if running) and permanently deletes all references to the H2O cluster. If it has been stopped with the model building state saved, this data will be permanently deleted as well.

Enterprise Steam as an H2O gateway

All H2O administration activities occur on Enterprise Steam, and users must launch H2O clusters through Steam. This *all roads lead to Enterprise Steam* approach means that Steam governs users and their H2O clusters before they are launched on the enterprise system. This is detailed in the following diagram:

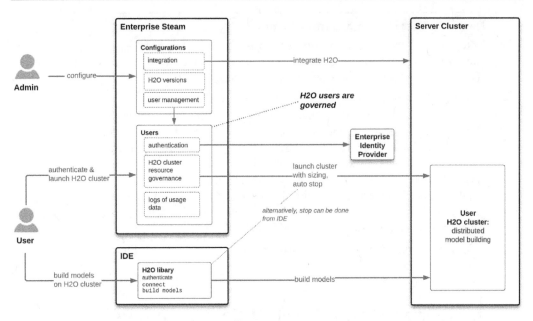

Figure 2.5 – Enterprise Steam viewed as an H2O gateway to the enterprise cluster

Administrators configure settings to manage H2O users and integrate Enterprise Steam with the enterprise server cluster. Additionally, administrators store H2O software versions that will be pushed to the server cluster when H2O clusters are launched and removed when the cluster is stopped and the resources are released. Administrators also have access to user usage data. This is all done through an administration-only UI.

Administrators configure users and how users launch H2O clusters in the enterprise environment. These configurations define limits on the number of concurrent clusters a user can launch simultaneously, the size (that is, the number of nodes), and the number of resources (for example, memory per node) allocated for each H2O cluster that is launched. Configurations also define when the cluster will stop or delete if the user does not do so manually from the H2O model building code in the IDE. A set of such configurations is defined as a profile, and one or more profiles are assigned to users or user groups. Therefore, administrators can assign some users as power users and others as light users.

Users authenticate to Enterprise Steam via the same identity provider (for example, LDAP) that was implemented to authorize access to resources on the enterprise server cluster environment (for example, S3 buckets). Enterprise Steam passes the user identity when the user launches a cluster, and this identity is used during authorization challenges on the enterprise system. Users in their IDEs must authenticate against the Enterprise Steam API to connect to the clusters they have launched.

Does H2O Core Require Enterprise Steam?

Note that H2O core does not require Enterprise Steam. Enterprise administrators can configure their enterprise server cluster infrastructure to allow H2O clusters to be launched on this infrastructure.

However, this approach is not a sound enterprise practice. It introduces a loss of control and governance that Enterprise Steam provides as a centralized H2O gateway to secure, manage, and log users, as elaborated in this section. Additionally, Enterprise Steam provides benefits to users by freeing them from the technical steps involved with integrating H2O Core with the enterprise cluster when launching H2O clusters, for example, Kerberos security requirements. The enterprise benefits of Enterprise Steam are explored in greater detail in *Chapter 11, The Administrator and Operations Views*, and in *Chapter 12, The Enterprise Architect and Security Views*.

Also, bear in mind that H2O Core is free and open source, whereas Enterprise Steam is not.

Enterprise Steam and the H2O Core high-level architecture

Now that we know how H2O clusters are formed and the role Enterprise Steam plays in administering H2O users and launching H2O clusters, let's understand Enterprise Steam and the H2O Core architecture from a high-level deployment perspective. The following diagram describes this deployment architecture:

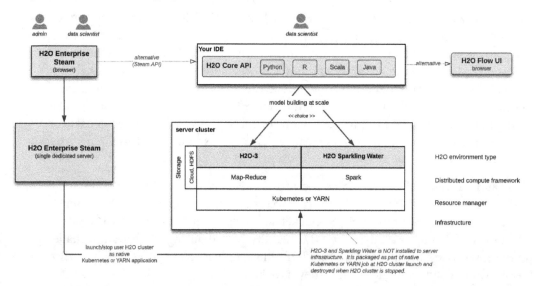

Figure 2.6 – Enterprise Steam and the H2O Core high-level deployment architecture

Enterprise Steam runs on its own dedicated server that communicates with the enterprise server cluster via HTTP(S). As mentioned earlier, Enterprise Steam stores the H2O Core (H2O-3 or Sparkling Water) JAR file that is pushed to the server cluster, which then self-organizes into a coordinated but distributed H2O cluster. This H2O cluster can be a native YARN or Kubernetes job, depending on which backend is implemented. Note that H2O-3 is run on a Map-Reduce framework, and Sparkling Water is run on the Spark framework.

An H2O-3 or Sparkling Water API library is installed to the data science IDE (for example, a `pip install` of the H2O-3 package in the Jupyter environment). It must match the version that is used to launch the cluster from Enterprise Steam. As mentioned previously, data scientists use the IDE to authenticate to Enterprise Steam, connect to the H2O cluster, and write H2O model building code. The H2O model building code is translated by the H2O client library into a REST message that is sent to the H2O cluster's leader node. Then, the work is distributed across the H2O cluster, and the results are returned to the IDE.

Note that enterprise clusters can be on-premise, cloud infrastructure-as-a-service, or managed service implementations. They can be, for example, Kubernetes or Cloudera CDH on-premise or in the cloud, or Cloudera CDP or Amazon EMR in the cloud. The full deployment possibilities are discussed in more detail in *Chapter 12, The Enterprise Architect and Security Views*.

H2O Platform Choices

H2O At Scale technology in this book is referred to as comprising: H2O Enterprise Steam + H2O Core (H2O-3, H2O Sparkling Water) + H2O MOJO. H2O At Scale integrates with an enterprise server cluster for model building and an enterprise scoring environment for model deployment.

H2O At Scale can be implemented with the just mentioned components alone. Alternatively, H2O At Scale can be implemented as a subset of the larger H2O machine learning platform and capability set called H2O AI Cloud. The H2O AI Cloud platform is described in greater detail in *Section 5 – Broadening the View – Data to AI Applications with the H2O AI Cloud Platform*.

Sparkling Water allows users to code in H2O and Spark seamlessly

The following code shows a simple example of Spark and H2O integrated in the same H2O code using H2O Sparkling Water:

```
# import data
loans_spark = spark.read.load("loans.csv", format="csv",
```

```
sep=",", inferSchema="true", header="true")


# Spark data engineering code
loans_spark = # any Spark SQL or Spark DataFrame code


# Convert Spark DataFrame to H2O Frame
loans = h2oContext.asH2OFrame(loans_spark)


# Continue with H2O model building steps as in previous code
example
loans.describe()
```

The code shows Spark importing data, which is held as a **Spark DataFrame**. **Spark SQL** or the **Spark DataFrame** API is used to engineer this data into a new DataFrame and then this Spark DataFrame is converted into an **H2OFrame** from which H2O model building is performed. Therefore, the user is iterating seamlessly from Spark to H2O code in the same API language and IDE.

The idea of the H2O cluster is still fundamentally true for Sparkling Water. It now expresses the H2O cluster architecture within the Spark framework. Details of this architecture are elaborated in *Chapter 12, The Enterprise Architect and Security Views*.

MOJOs export as DevOps-friendly artifacts

Data scientists build models, but the end goal is to put models into a production environment where predictions are made in a business context. MOJOs make this last mile of deployment easy. MOJOs are exported by a single line of code. For example, whether the model was built using Python, R, or using a generalized linear model, an XGBoost model, or stacked ensemble, all MOJOs are identical from a DevOps perspective. This makes model deployment repeatable and, thus, capable of fitting into existing automated CI/CD pipelines that are used throughout the organization.

Summary

In this chapter, we laid the foundation for understanding H2O machine learning at scale. We started by reviewing a bare minimum *Hello World* code example and discussed the problems of scale around it. Then, we introduced the H2O Core, Enterprise Steam, and MOJO technology components and how these can overcome problems of scale. Finally, we extracted a set of key concepts from these technologies to deepen our understanding.

In the next chapter, we will use this understanding to begin our journey of learning how to build and deploy world-class models at scale. Let the coding begin!

3
Fundamental Workflow – Data to Deployable Model

In this chapter, we will walk through a minimal model-building workflow for H2O at scale. We will refer to this as the *fundamental workflow* because it omits the wide range of functionality and user choices to build accurate, trusted models while nevertheless touching on the main steps.

The fundamental workflow will serve as a basis to build your understanding of H2O technology and coding steps so that in the next part of the book you can dive fully into advanced techniques to build state-of-the-art models.

To develop the fundamental workflow, we will cover the following main topics in this chapter:

- Use case and data overview
- The fundamental workflow
- Variation points – alternatives and extensions to the fundamental workflow

Technical requirements

For this chapter, we will focus on using Enterprise Steam to launch H2O clusters on an enterprise server cluster. Enterprise Steam technically is not required to launch H2O clusters but enterprise stakeholders typically view Enterprise Steam as a security, governance, and administrator requirement for implementing H2O in enterprise environments.

Enterprise Steam requires a license purchased from H2O.ai. If your organization does not have an instance of Enterprise Steam installed, you can access Enterprise Steam and an enterprise server cluster through a temporary trial license of the larger H2O platform. Alternatively, for ease of conducting the exercises in this book, you may wish to launch H2O clusters as a sandbox in your local environment (for example, on your laptop or desktop workstation) and bypass the use of Enterprise Steam.

See *Appendix – Alternative Methods to Launch H2O Clusters for this Book* to help you decide on how you wish to launch H2O clusters for the exercises in this book and how to set up your environment to do so.

> **Enterprise Steam: Enterprise Environment versus Coding Exercises in the Book**
>
> Enterprise stakeholders typically view Enterprise Steam as a security, governance, and administrator requirement for implementing H2O in enterprise environments. This chapter shows how data scientists use Enterprise Steam in this enterprise context. Enterprise Steam, however, requires an H2O.ai license to implement and will not be available to all readers of this book.
>
> A simple sandbox (non-enterprise) experience is to use H2O exclusively on your local environment (laptop or workstation) and this does not require Enterprise Steam. Coding exercises in subsequent chapters will leverage the local sandbox environment but also can be performed using Enterprise Steam as demonstrated in this chapter.
>
> Note that the distinction between the data scientist workflow with and without Enterprise Steam is isolated to the first step of the workflow (launching the H2O cluster) and will be made clearer later in this chapter. See also *Appendix – Alternative Methods to Launch H2O Clusters*.

Use case and data overview

To demonstrate the fundamental workflow, we will implement a binary classification problem where we predict the likelihood that a loan will default or not. The dataset we use in this chapter can be found at `https://github.com/PacktPublishing/Machine-Learning-at-Scale-with-H2O/blob/main/chapt3/loans-lite.csv`. (This is a simplified version of the Kaggle *Lending Club Loan* dataset: `https://www.kaggle.com/imsparsh/lending-club-loan-dataset-2007-2011`.)

We are using a simplified version of the dataset to streamline the workflow in this chapter. In *Part 2, Building State-of-the-Art Models at Scale*, we will develop this use case using advanced H2O model-building capabilities on the original loan dataset.

The fundamental workflow

Our fundamental workflow will proceed through the following steps:

1. Launching the H2O cluster (Enterprise Steam UI)
2. Connecting to the H2O cluster (your IDE from this point onward)
3. Building the model
4. Evaluating and explaining the model
5. Exporting the model for production deployment
6. Shutting down the H2O cluster

Step 1 – launching the H2O cluster

This step is done from the Enterprise Steam UI. You will select whether you want an H2O-3 or Sparkling Water cluster and then you will configure the H2O cluster behavior, such as the duration of idle time before it times out and terminates and whether you want to save the state at termination so you can restart the cluster and pick up where you left off (this must be enabled by the administrator). Nicely, Enterprise Steam will auto-size the H2O cluster (number of nodes, memory per node, CPUs) based on your data size.

Logging in to Steam

Open a web browser and go to `https://steam-url:9555/login` and log in to Enterprise Steam, where `steam-url` is the URL of your specific Steam instance. (Your administrator may have changed the port number, but typically it is `9555` as shown in the URL.)

Selecting an H2O for H2O-3 (versus Sparkling Water) cluster

Here, we will launch an H2O-3 cluster (and not Sparkling Water, which we will do in the next part of the book), so click on the **H2O** link in the left panel and then click **LAUNCH NEW CLUSTER**.

Configuring the H2O-3 cluster

This brings us to the following form, which you will configure:

Home > H2O > Clusters > Launch

NEW H2O CLUSTER

SELECT PROFILE	default-h2o-kubernetes ⬍		
CLUSTER NAME		required must meet requirements	❷
H2O VERSION	3.32.0.3 ⬍	required	❷
DATASET PARAMETERS	Set parameters 🔁 (applied)	optional	❷
NUMBER OF NODES	6	required min: 1, max: 8	❷
NUMBER OF CPUS	8	required min: 1, max: 8	❷
NUMBER OF GPUS	0	required min: 0, max: 0	❷
MEMORY PER NODE [GB]	48	required min: 1, max: 48	❷
MAXIMUM IDLE TIME [HRS]	8	required min: 1, max: 24	❷
MAXIMUM UPTIME [HRS]	12	required min: 1, max: 24	❷
TIMEOUT [S]	600	required min: 60, max: 1800	❷

Figure 3.1 – UI to launch an H2O-3 cluster on Kubernetes

For now, we will ignore most configurations. These will be covered more fully in *Chapter 11*, *The Administrator and Operations Views*, where Enterprise Steam is overviewed in detail. Note that the configuration page uses the term *H2O cluster* to represent an H2O-3 cluster specifically, whereas in this book we use the term H2O cluster to represent either an H2O-3 or Sparkling Water cluster.

> **Note on the "Configuring the H2O-3 cluster" Screenshot**
>
> Details on the screen shown in *Figure 3.1* will vary depending on whether the H2O cluster is launched on a Kubernetes environment or on a YARN-based Hadoop or Spark environment. Details will also vary based on whether the H2O cluster is an H2O-3 cluster or a Sparkling Water cluster. In all cases, however, the fundamental concepts of H2O cluster size (number of nodes, CPU/GPU per node, and memory per node) and maximum idle/uptime are common throughout.

Give your cluster a name and for **DATASET PARAMETERS**, click **Set parameters** to arrive at the following popup:

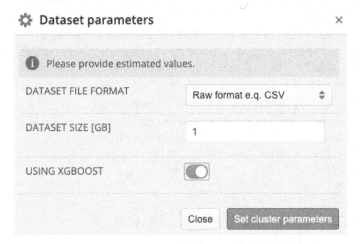

Figure 3.2 – Popup to automatically size the H2O-3 cluster

The inputs here are used by Enterprise Steam to auto-size your H2O cluster (that is, to determine the number of H2O nodes and memory allocated for each node and CPU allocations for each node). Recall the *key concepts* of an H2O cluster as presented in the previous chapter.

Waiting briefly for the cluster to start

The **STATUS** field in the UI will state **Starting**, signifying that the H2O cluster is being launched on the enterprise server cluster. This will take a minute or two. When the status changes to **Running**, your H2O cluster is ready to use.

Viewing details of the cluster

Let's first learn a few things about the cluster by clicking on **Actions** and then **Detail**. This generates a popup describing the cluster.

Notice in this case that **Number of nodes** is **6** and **Memory per node** is **48 GB** as auto-sized by Enterprise Steam for a dataset size of 50 GB, as shown in *Figure 3.1*. Recall from the *H2O key concepts* section in the previous chapter that our dataset is partitioned and distributed in memory across this number of H2O cluster nodes on the enterprise server cluster and that compute is done in parallel on these H2O nodes.

Note on H2O Cluster Sizing

In general, an H2O cluster is sized so the total memory allocated to the cluster (that is, the product of N H2O nodes and X GB memory per node) is roughly 5 times the size of the uncompressed dataset that will be used for model building. The calculation minimizes the number of nodes (that is, fewer nodes with more memory per node is better).

Enterprise Steam will calculate this sizing based on your description of the dataset, but alternatively, you can size the cluster yourself through the Enterprise Steam UI. The total memory allocated to the H2O cluster will be released when the H2O cluster is terminated.

Note that the Enterprise Steam administrator sets the minimum and maximum configuration values a user may have when launching an H2O cluster (see *Figure 3.1*) and thus the maximum H2O cluster size a user may launch. These boundaries set by the administrator can be configured differently for different users.

Step 2 – connecting to the H2O cluster

This and all subsequent steps are from your IDE. We will use a Jupyter notebook and write code in Python (though other options include writing H2O in R, Java, or Scala using your preferred IDE).

Open the notebook and connect to the H2O cluster you launched in Enterprise Steam by writing the following code:

```
import h2o
import h2osteam
from h2osteam.clients import H2oKubernetesClient
conn = h2osteam.login(url="https://steam-url:9555",
                      username="my-steam-username",
                      password="my-steam-password")
cluster = H2oKubernetesClient().get_cluster("cluster-name")
cluster.connect()
```

You have now connected to the H2O cluster and can start building models. Note that after you connect, you will see H2O cluster details similar to those viewed from the Enterprise Steam UI when you configured the cluster before launching.

Let's understand what the code is doing:

1. You referenced the `h2osteam` and `h2o` Python libraries that were downloaded from H2O and implemented in the IDE environment. (The `h2o` library is not used by the code shown here but will be used by subsequent model building steps that follow.)

2. Then you logged into the Enterprise Steam server via the `h2osteam` API (library). You used the same URL, username, and password that was used to log in to the UI of Enterprise Steam.

3. You then retrieved your H2O cluster information from Enterprise Steam via the `h2osteam` API.

4. Note that you are using `H2oKubernetesClient` here because you are connecting to an H2O cluster launched on a Kubernetes environment. If, alternatively, your enterprise environment is Hadoop or Spark, you use `H2oClient` or `SparklingClient`, respectively.

5. You connected to your H2O cluster using `cluster.connect()` and passed the cluster information to the `h2o` API. Note that you did not have to specify any URL to the H2O cluster because Steam returned this behind the scenes with `H2oKubernetesClient().get_cluster("cluster-name")`.

Creating an H2O Sandbox Environment

If you want to create a small H2O sandbox on your local machine instead of using Enterprise Steam and your enterprise server cluster, simply implement the following two lines of code from your IDE:

```
import h2o

h2o.init()
```

The result is identical to performing *steps 1–2* using Enterprise Steam, except that it launches an H2O cluster with one node on your local machine and connects to it.

Whether connecting to an H2O cluster in your enterprise environment or on your local machine, you can now write model-building steps identically from your IDE against the respective cluster. For the sandbox, you will be constrained, of course, to much smaller data volumes because of its small cluster size of one node with low memory.

Step 3 – building the model

Now that we have connected to our H2O cluster, it is time to build the model. From this point onward, you will be using the h2o API to communicate with the H2O cluster to which you launched and connected.

Here in our fundamental workflow, we will take a minimal approach to import data, clean it, engineer features from it, and then train the model.

Importing the data

The loans dataset is loaded from the source into the H2O-3 cluster memory using the h2o.import_file command as follows:

```
input_csv = "https://raw.githubusercontent.com/PacktPublishing/
Machine-Learning-at-Scale-with-H2O/main/chapt3/loans-lite.csv"
loans = h2o.import_file(input_csv)
loans.dim
loans.head()
```

The loans.dim line gives us the number of rows and columns and loans.head() displays the first 10 rows. Quite simple data exploration for now.

Note that the dataset is now partitioned and distributed in memory across the H2O cluster. From our coding standpoint in the IDE, it is treated as a single two-dimensional data structure of columns and rows called an **H2OFrame**.

Cleaning the data

Let's perform one simple data cleaning step. The target or response column is called bad_loan and it holds values of either 0 or 1 for good and bad loans respectively. We need to transform the integers in this column to categorical values, as shown next:

```
loans["bad_loan"] = loans["bad_loan"].asfactor()
```

Engineering new features from the original data

Feature engineering is often considered the *secret sauce* in building a superior predictive model. For our purposes now, we will do basic feature engineering by extracting year and month as separate features from the issue_d column, which holds day, month, and year as a single value:

```
loans["issue_d_year"] = loans["issue_d"].year().asfactor()
loans["issue_d_month"] = loans["issue_d"].month().asfactor()
```

We have just created two new categorical columns in our `loans` dataset: `issue_d_year` and `issue_d_month`.

Model training

We will next train a model to predict bad loans. We first split our data into `train` and `test`:

```
train, validate, test = loans.split_frame(seed=1, ratios=[0.7,
0.15])
```

We now need to identify which columns we will use to predict whether a loan is bad or not. We will do this by removing two columns from the current loans H2OFrame, which hold the cleaned and engineered data:

```
predictors = list(loans.col_names)
predictors.remove("bad_loan)
predictors.remove("issue_d")
```

Note that we removed `bad_loan` from the columns used as features because this is what we are predicting. We also removed `issue_d` because we engineered new features from this and do not want it as a predictor.

Next, let's create an XGBoost model to predict loan default:

```
from h2o.estimators import H2OXGBoostEstimator
param = {
            "ntrees" : 20,
            "nfolds" : 5,
            "seed": 12345
}
model = H2OXGBoostEstimator(**param)
model.train(x = predictors,
            y = "bad_loan",
            training_frame = train,
            validation_frame = validate)
```

Step 4 – evaluating and explaining the model

Let's evaluate the performance of the model that we just trained:

```
perf = model.model_performance(test)
perf
```

The output of `perf` shows details on model performance, including model metrics such as MSE, Logloss, AUC, and others, as well as a confusion matrix, maximum metrics thresholds, and a gains/lift table.

Now let's look at one simple view of model explainability by generating variable importance from the model result:

```
explain = model.explain(test,include_explanations="varimp")
explain
```

The output of `explain` shows the variable importance of the trained model run against the test dataset. This is a table listing how strongly each feature contributed to the model.

H2O's model explainability capabilities go much further than variable importance, as we shall see later in the book.

Step 5 – exporting the model's scoring artifact

Now let's generate and export the model as a scoring artifact that can be deployed to a production environment by the DevOps group:

```
model.download_mojo("download-destination-path")
```

In the real world, of course, we would train many models and compare their performance and explainability to evaluate which (if any) should make it to production.

Step 6 – shutting down the cluster

When your work is complete, shut down the H2O-3 cluster to free up the resources that were reserved by it:

```
h2o.cluster().shutdown()
```

Variation points – alternatives and extensions to the fundamental workflow

The fundamental workflow we developed here is a simple example. For each step we performed, there are multiple alternatives and extensions to what has been shown. All of *Part 2:, Building State-of-the-Art Models at Scale,* is dedicated to understanding these alternatives and elaborations and to putting them together to build superior models at scale.

Let's first touch on some key variation points here.

Launching an H2O cluster using the Enterprise Steam API versus the UI (step 1)

In our example, we used the convenience of the Enterprise Steam UI to configure and launch an H2O cluster. Alternatively, we could have used the Steam API from our IDE to do so. See the full H2O Enterprise Steam API documentation at `https://docs.h2o.ai/enterprise-steam/latest-stable/docs/python-docs/index.html` for the Python API and `https://docs.h2o.ai/enterprise-steam/latest-stable/docs/r-docs/index.html` for the R API.

By launching the H2O cluster from our IDE, we therefore could have completed all of *steps 1–6* of our workflow exclusively from the IDE.

Launching an H2O-3 versus Sparkling Water cluster (step 1)

In our example, we launched an H2O-3 cluster. We could alternatively launch an H2O Sparkling Water cluster. As we will see, Sparkling Water clusters have the same capability set as H2O-3 clusters but with the additional ability to integrate Spark code and Spark DataFrames with H2O code and H2O DataFrames. This is particularly powerful when leveraging Spark for advanced data exploration and data munging before building models in H2O.

Implementing Enterprise Steam or not (steps 1–2)

Know that Enterprise Steam is not a requirement for launching and connecting to an enterprise server cluster: it is possible for a data scientist to use only the h2o (and not h2osteam) API in the IDE to configure, launch, and connect to an enterprise server cluster, but this is low-level coding and configuration and requires detailed integration information. Importantly, this approach lacks sound enterprise security, governance, and integration practices.

In the enterprise setting, Enterprise Steam is viewed as essential to centralize, manage, and govern H2O technology and H2O users in the enterprise server cluster environment. These capabilities are elaborated on in *Chapter 11, The Administrator and Operations Views.*

Using a personal access token to log in to Enterprise Steam (step 2)

For *Step 2 – connecting to the H2O cluster*, we authenticated to Enterprise Steam from our IDE using the Enterprise Steam API. In the example code, we used a clear text password (which was the same password used to log into the Enterprise Steam UI). This is not secure if, for example, you shared the notebook.

Alternatively, and more securely, you can use a **Personal Access Token** (**PAT**) as the API login password to Enterprise Steam. A PAT can be generated as often as you wish, with each newly generated PAT revoking the previous one. Thus, if you shared a Jupyter notebook with your login credentials using a PAT as your password, the recipient of the notebook would not know your Enterprise Steam UI login password and could not authenticate via the API using the revoked password in your shared notebook. You can take the PAT one step further and implement it as an environment variable outside the IDE.

Enterprise Steam lets you generate a PAT from the UI. To generate a PAT, log in to Enterprise Steam UI, click **Configurations,** and follow the brief token workflow. Copy the result (a long string) for use in your current notebook or script or to set it as an environment variable.

Building the model (step 3)

H2O offers a much more powerful model-building experience than what was shown in our fundamental workflow. This larger experience is touched on here and explored fully in *Part 2, Building State-of-the-Art Models at Scale.*

Language and IDE

We are writing H2O code in Python in a Jupyter notebook. You can also choose R for the Enterprise Steam API and use the Python or R IDE of your choice. Additionally, you can use H2O's UI-rich IDE called **H2O Flow** to perform the full workflow or to quickly understand aspects of an H2O cluster workflow that is progressing from your own IDE.

Importing data

Data can be imported from many sources into H2O clusters, including cloud object storage (for example, S3 or Azure Delta Lake), database tables (via JDBC), HDFS, and more. Additionally, source files can have many formats, including Parquet, ORC, ARFF, and more.

Cleaning data and engineering features

H2O-3 has capabilities for basic data manipulation (for example, changing column types, combining or slicing rows or columns, group by, impute, and so on).

Recall that launching a Sparkling Water cluster gives us full H2O-3 capabilities with the addition of Spark's more powerful data exploration and engineering capabilities.

Model training

In our fundamental workflow, we explored only one type of model (XGBoost) while changing only a few default parameters. H2O-3 (and its Sparkling Water extension) has an extensive list of both supervised and unsupervised learning algorithms and a wide range of parameters and hyperparameters to set to your specification. In addition, these algorithms can be combined powerfully into an AutoML workflow that explores multiple models and hyperparameter space and arranges the resulting best models on a leaderboard. You also have control over cross-validation techniques, checkpointing, retraining, and reproducibility.

Evaluating and explaining the model (step 4)

H2O has numerous explainability methods and visualizations for both local (individual) and global (model-level) explainability, including residual analysis, variable importance heatmaps, Shapley summaries, **Partial Dependence Plots** (**PDPs**), and **Individual Conditional Expectation** (**ICE**).

Exporting the model's scoring artifact (step 5)

Once you export the model's scoring artifact (called an H2O MOJO), it is ready for DevOps to deploy and monitor in live scoring environments. It likely will enter the organization's CI/CD process. We will pick it up at this point in *Part 3, Deploying Your Models to Production Environments*.

Shutting down the cluster (step 6)

You can shut down your cluster from your IDE as shown in our example workflow. If you noticed, when configuring your cluster in Enterprise Steam, however, there are two configurations that automate the shutdown process: **MAXIMUM IDLE TIME** and **MAXIMUM UPTIME**. The first shuts down the cluster after it has not been used for the configured amount of time. The second shuts down the cluster after it has been up for the configured amount of time. Shutting down clusters (manually or automatically) saves resources for others using the enterprise server cluster.

The administrator assigns minimum and maximum values for these auto-terminate configurations. Note that when enabled by administrators, Enterprise Steam saves all models and DataFrames when the H2O cluster has been auto-terminated. You can restart the cluster later and pick up where the cluster terminated.

Summary

In this chapter, you learned how to launch an H2O cluster and build a model on it from your IDE. This fundamental workflow is a bare skeleton that you will flesh out much more fully with a deep set of advanced H2O model-building techniques that we will now learn in *Part 2, Building State-of-the-Art Models at Scale,* of the book.

We will start this advanced journey in the next chapter by overviewing these capabilities before using them.

Section 2 – Building State-of-the-Art Models on Large Data Volumes Using H2O

This section dives deep into advanced techniques to build accurate and trusted ML models with large to massive data volumes using H2O. We first overview the full capability set of H2O-3 and Sparkling Water for model building. From there, we demonstrate these capabilities by engineering features, building and optimizing supervised learning models, building H2O models embedded in Spark pipelines, building unsupervised models using H2O algorithms, and reviewing how to update and ensure the reproducibility of H2O models. From there, we introduce in depth a number of methods for interpreting and understanding the decision-making process of your model and introduce auto-documentation within H2O. Finally, we do an extensive and thorough exercise in model building from problem statement and raw data through data cleaning, feature engineering, model building and optimization, and candidate model selection based on performance and explainability.

This section comprises the following chapters:

- *Chapter 4, H2O Model Building at Scale – Capability Articulation*
- *Chapter 5, Advanced Model Building – Part I*
- *Chapter 6, Advanced Model Building – Part II*
- *Chapter 7, Understanding ML Models*
- *Chapter 8, Putting It All Together*

4
H2O Model Building at Scale – Capability Articulation

So far, we have learned the fundamental workflow of how to build H2O models at scale, but that was done using H2O at its barest minimum. In this chapter, we will survey the extremely broad capability set of H2O model building at scale. We will then use our knowledge from this chapter and move on to part two, *Building State-of-the-Art Models on Large Data Volumes Using H2O*, where we will get down to business and use advanced techniques to build and explain highly predictive models at scale.

To conduct this survey, we will break down the chapter into the following main topics:

- Articulating the H2O data capabilities during model building
- Overviewing the H2O machine learning algorithms
- Understanding the H2O modeling capabilities

H2O data capabilities during model building

Recall that H2O model building at scale is performed by using H2O 3 or its extension, Sparkling Water, which wraps H2O 3 with Spark capabilities. The H2O 3 API has extensive data capabilities used in the model building process, and the Sparkling Water API inherits these and adds additional capabilities from Spark. These capabilities are broken down into the following three broad categories:

- **Ingesting data** from the source to the H2O cluster
- **Manipulating data** on the H2O cluster
- **Exporting data** from the H2O cluster to an external destination

As emphasized in previous chapters, the H2O cluster architecture (H2O 3 or Sparkling Water) allows model building at an unlimited scale but is abstracted from the data scientist who builds models by coding H2O in the IDE.

H2O data capabilities are overviewed in the following diagram and elaborated subsequently:

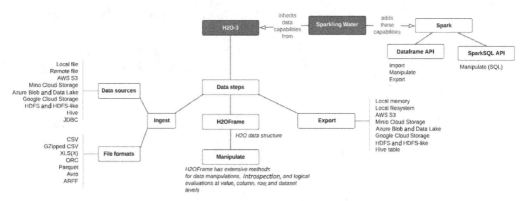

Figure 4.1 – The H2O data capabilities

Let's start with data ingestion.

Ingesting data from the source to the H2O cluster

The following data sources are supported:

- **Local file**
- **Remote file**
- **AWS S3**
- **MinIO cloud storage**

- **Azure Blob and Data Lake**
- **Google Cloud Storage**
- **HDFS**
- **HDFS-like**: Alluxio FS and IBM HDFS
- **Hive** (via Metastore/HDFS or JDBC)
- **JDBC**

The supported file formats of source data are as follows:

- **CSV** (a file with any delimiter, auto-detected or specified)
- **GZipped CSV**
- **XLS** or **XLSX**
- **ORC**
- **Parquet**
- **Avro**
- **ARFF**
- **SVMLight**

Some important characteristics of data ingestion to H2O are as follows:

- Data is ingested directly from the source to the H2O cluster memory and does not pass through the IDE client.
- In all cases, data is partitioned in-memory across the H2O cluster.
- Except for the local file, the remote file, and JDBC sources, data is ingested in parallel to each partition.
- Data on the H2O cluster is represented to the user in the IDE as a two-dimensional **H2OFrame**.

Let's now see how we can manipulate data now that it is ingested into H2O and represented as an H2OFrame.

Manipulating data in the H2O cluster

The H2O 3 API provides extensive data manipulation capabilities. As mentioned in the previous bullet list, datasets in memory are distributed on the H2O cluster and represented in the IDE specifically as an H2OFrame after data load and subsequent data manipulations.

H2OFrames have an extensive list of methods to perform mathematical, logical, and introspection operations at the value, column, row, and full dataset levels. An H2OFrame is similar in experience to the **pandas DataFrame** or **R data frame**.

The following examples are just a few data manipulations that can be done on H2Oframes:

- Operations on **data columns**:

  - Change the data type (for example, integers from 0 to 7 as categorical values).

  - Aggregate a column (group by) by applying mathematical functions.

  - Display column names and use as features in a model.

- Operations on **data rows**:

  - Combine rows from one or more datasets.

  - Slice out (filter) rows of a dataset by specifying the row index, the range of rows, or the logical condition.

- Operations on **datasets**:

  - Merge two datasets on common values of shared column names.

  - Transform a dataset by pivoting on a column.

  - Split a dataset into two or more datasets (for example, train, validate, and test).

- Operations on **data values**:

  - Fill missing values forward or backward with adjacent row or column values.

  - Fill missing values by imputing with aggregate results (for example, the mean for the column).

  - Replace numerical values based on logical conditions.

  - Trim values, manipulate strings, return a numerical value sign, and test whether a value is N/A.

- **Feature engineering** operations:

  - Date parsing, for example, parsing one date column into separate columns for year, month, day.

  - Derive a new column mathematically and conditionally from other columns, including the use of lambda expressions.

  - Perform target encoding (that is, replace a categorical value with the mean of the target variable).

  - For **Natural Language Processing (NLP)** problems, perform **string tokenizing**, **Term Frequency-Inverse Document Frequency (TF-IDF)** calculations, and convert a **Word2vec** model into an H2OFrame for data manipulations.

For full details of H2O data manipulation possibilities, see the H2O Python documentation (`http://docs.h2o.ai/h2o/latest-stable/h2o-py/docs/frame.html`) or R documentation (`http://docs.h2o.ai/h2o/latest-stable/h2o-r/docs/reference/index.html`). Also, refer to the fourth section of *Machine Learning with Python and H2O* (`http://h2o-release.s3.amazonaws.com/h2o/rel-wheeler/2/docs-website/h2o-docs/booklets/PythonBooklet.pdf`) for examples of data manipulation.

Manipulating data is key for preparing it as an input for model building. We may also want to export our manipulated data for future use. The next section lists the H2O data export capabilities.

Exporting data out of the H2O cluster

H2OFrames in memory can be exported to external targets. These target systems are as follows:

- **Local client memory**
- **Local filesystem**
- **AWS S3**
- **MinIO cloud storage**
- **Azure Blob and Data Lake**

- **Google Cloud Storage**
- **HDFS**
- **HDFS-like**: Alluxio FS and IBM HDFS
- **Hive tables** (CSV or Parquet, via JDBC)

The volume of exported data must, of course, be considered. Large volumes of data will not, for example, fit into a local client memory or filesystem.

Let's now see what additional data capabilities Sparkling Water adds.

Additional data capabilities provided by Sparkling Water

Sparkling Water inherits all data capabilities from H2O 3. Importantly, Sparkling Water adds additional data capabilities by leveraging the Spark DataFrame and Spark SQL APIs, and thus can import, manipulate, and export data accordingly. See the following reference for full Spark DataFrame and Spark SQL capabilities: `https://spark.apache.org/docs/latest/sql-programming-guide.html`.

A key pattern in using Sparkling Water is to leverage Spark for advanced data munging capabilities, then convert the resulting Spark DataFrame to an H2Oframe, and then build state-of-the-art models using H2O's machine learning algorithms, as covered in the next section. These algorithms can be used in either H2O 3 or Sparkling Water.

H2O machine learning algorithms

H2O has extensive **unsupervised** and **supervised** learning algorithms with similar reusable API constructs – for example, similar ways to set hyperparameters or invoke explainability capabilities. These algorithms are identical from an H2O 3 or Sparkling Water perspective and are overviewed in the following diagram:

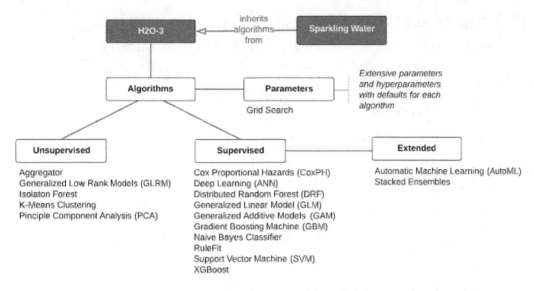

Figure 4.2 – H2O algorithms

Each algorithm has an extensive set of parameters and hyperparameters to set or leverage as defaults. The algorithms accept H2OFrames as data inputs. Remember that an H2OFrame is simply a handle on the IDE client to the distributed in-memory data on the remote H2O cluster where the algorithm processes it.

Let's take a look at H2O's distributed machine learning algorithms.

H2O unsupervised learning algorithms

Unsupervised algorithms do not predict but rather attempt to find clusters and anomalies in data, or to reduce the dimensionality of a dataset. H2O has the following unsupervised learning algorithms to run at scale:

- **Aggregator**
- **Generalized Low Rank Models (GLRM)**
- **Isolation Forest**
- **Extended Isolation Forest**
- **K-Means Clustering**
- **Principal Component Analysis (PCA)**

H2O supervised learning algorithms

Supervised learning algorithms predict outcomes by learning from a training dataset labeled with those outcomes. H2O has the following supervised learning algorithms to run at scale:

- **Cox Proportional Hazards (CoxPH)**
- **Deep Learning (Artificial Neural Network, or ANN)**
- **Distributed Random Forest (DRF)**
- **Generalized Linear Model (GLM)**
- **Maximum R Square Improvements (MAXR)**
- **Generalized Additive Models (GAM)**
- **ANOVA GLM**
- **Gradient Boosting Machine (GBM)**
- **Naïve Bayes Classifier**
- **RuleFit**
- **Support Vector Machine (SVM)**
- **XGBoost**

Parameters and hyperparameters

Each algorithm has a deep set of parameters and hyperparameters for configuration and tuning. Specifying most parameters is optional; if not specified, the default will be used. Parameters include the specification of cross-validation parameters, learning rates, tree depths, weights columns, ignored columns, early stopping parameters, the distribution of response column (for example, Bernoulli), categorical encoding schemes, and many other specifications.

You can dive deeper into H2O's algorithms and their parameters in H2O's documentation at `http://docs.h2o.ai/h2o/latest-stable/h2o-docs/data-science.html#algorithms`. The H2O website also lists tutorials and booklets for its algorithms at `http://docs.h2o.ai/#h2o`. A full list of algorithm parameters, each with a description, status as a hyperparameter or not, and mapping to algorithms that use the parameter, are found in H2O's documentation appendix at `http://docs.h2o.ai/h2o/latest-stable/h2o-docs/parameters.html`.

H2O extensions of supervised learning

H2O extends its supervised learning algorithms by providing **Automatic Machine Learning (AutoML)** and **Stacked Ensemble** capabilities. We will take a closer look` at these in the next section, where we will place H2O algorithms in the larger context of model capabilities.

Miscellaneous

H2O provides utilities to enhance work with its algorithms. **Target encoding** helps you handle categorical values and has many configurable parameters to make this easy. **TF-IDF** and **Word2vec** are commonly used in NLP problems, and they also are nicely configurable. Finally, **permutation variable importance** is a method to help understand how strongly your features contribute to the model and can help in evaluating which features to use in your final training dataset.

H2O modeling capabilities

H2O's supervised learning algorithms are used to train models on training data, tune them on validation data, and score or predict with them on test or live production data. H2O has extensive capabilities to train, evaluate, explain, score, and inspect models. These are summarized in the following diagram:

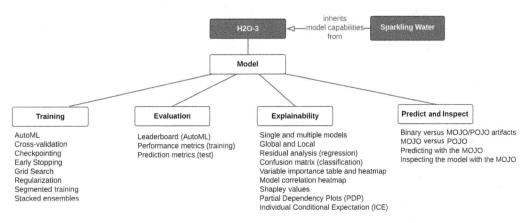

Figure 4.3 – The H2O supervised learning capabilities

Let's take a closer look at the model training capabilities.

H2O model training capabilities

Algorithms are at the heart of model training, but there are a larger set of capabilities to consider beyond the algorithms themselves. H2O provides the following model training capabilities:

- **AutoML**: An easy-to-use interface and parameter set that automates the process of training and tuning many different models, using multiple algorithms, to create a large number of models in a short amount of time.

- **Cross-validation**: K-fold validation is used to generate performance metrics against folds of the validation split, and parameters such as the number of folds can be specified in the algorithm's training parameters.

- **Checkpointing**: A new model is built as a continuation from a previously trained, checkpointed model as opposed to building the model from scratch; this is useful, for example, in retraining a model with new data.

- **Early stopping**: The parameters to define when an algorithm stops model building early, determined by which of many stopping metrics is specified.

- **Grid search**: Build models for each combination of a range of hyperparameters that are specified and sort the resulting models by a performance metric.

- **Regularization**: Most algorithms have parameter settings to specify regularization techniques to prevent overfitting and increase explainability.

- **Segmented training**: Training data is partitioned into segments based on the same column values, and a separate model is built for each segment.

- **Stacked ensembles**: Combines the results of multiple base models that use the same or different algorithms into a better-performing single model.

After training a model, we want to evaluate it to determine whether its predictive performance meets our needs. Let's see what H2O offers in this regard.

H2O model evaluation capabilities

H2O exposes many model attributes to evaluate model performance. These are summarized as follows:

- The **leaderboard for AutoML**: The AutoML model results ranked by configured performance metrics or other attributes, such as average prediction speed, with additional metrics shown.

- **Performance metrics for classification problems**: For classification problems, H2O calculates **GINI coefficient, Absolute Matthew Correlation Coefficient (MCC), F1, F0.5, F2, Accuracy, Logloss, Area Under the ROC Curve (AUC), Area Under the Precision-Recall Curve (AUCPR)**, and **Kolmogorov-Smirnov (KS)** metrics.

- **Performance metrics for regression problems**: For regression problems, H2O calculates **R Squared (R²), Mean Squared Error (MSE), Root Mean Squared Error (RMSE), Root Mean Squared Logarithmic Error (RMSLE)**, and **Mean Absolute Error (MAE)** metrics.

- **Prediction metrics**: After a model is built, H2O allows you to predict a leaf node assignment (tree-based models), feature contributions, class probabilities for each stage (GBM models), and feature frequencies on a prediction path (GBM and DRF).

- **Learning curve plot**: This shows a model performance metric as learning progresses to help diagnose overfitting or underfitting.

Let's now explore ways to explain H2O models.

H2O model explainability capabilities

H2O presents a simple and uniform interface to explain either single models or multiple models, which can be a list of separately built models or a reference to those generated from AutoML. On top of that, H2O allows you to generate global (that is, model-level) and local (row- or individual-level) explanations. H2O's explainability capabilities are configurable to your specifications. Output is tabular, graphical, or both, depending on the explanation.

We will dedicate all of *Chapter 6, Advanced Model Building – Part II*, to explore this important topic in greater detail, but for now, here is a quick list of capabilities:

- **Residual analysis for regression**

- **Confusion matrix for classification**

- **Variable importance table and heatmap**

- **Model correlation heatmap**

- **Shapley values**

- **Partial Dependency Plots (PDPs)**

- **Individual Conditional Expectation (ICE)**

Let's now complete our survey of H2O's capabilities for modeling at scale by seeing what we can do once our model is trained, evaluated, and explained.

H2O trained model artifacts

Once a model is trained, it can be exported and saved as a scoring artifact. The larger topic of deploying your artifact for production scoring will be treated in *Part 3: Deploying Your Model to Production Environments*. Here are the fundamental capabilities of the exported scoring artifact:

- **Predicting with a MOJO**: Models can be saved as self-contained binary Java objects called MOJOs that can be flexibly implemented as low-latency production scoring artifacts on diverse systems (for example, a REST server, batch database scoring, and Hive UDFs). MOJOs also can be reimported into H2O clusters for purposes described in the next bullet point.

- **Inspecting the model with a MOJO**: An exported MOJO can be re-imported into the H2O cluster and used to score against a dataset, inspect hyperparameters used to train the original model, see the scoring history, and show feature importances.

- **A MOJO compared to a POJO**: The POJO is the precursor to the MOJO and is being deprecated by H2O but is still required for some algorithms.

Summary

In this chapter, we conducted a wide survey of H2O capabilities for model building at scale. We learned about the data sources we can ingest into our H2O clusters and the file formats that are supported. We learned how this data moves from the source to the H2O cluster, and how the H2OFrame API provides a single handle in the IDE to represent the distributed in-memory data on the H2O cluster as a single two-dimensional data structure. We then learned the many ways in which we can manipulate data through the H2OFrame API and how to export it to external systems if need be.

We then surveyed the core of H2O model building at scale – H2O's many state-of-the-art distributed unsupervised and supervised learning algorithms. Then, we put those into context by surveying model capabilities around them, from training, evaluating, and explaining the models, to using model artifacts to retrain, score and inspect models.

With this map of the landscape firmly in hand, we can now roll up our sleeves and start building state-of-the-art H2O models at scale. In the next chapter, we will start by implementing the advanced model building topics one by one, before later putting it all together in a fully developed use case.

5
Advanced Model Building – Part I

In this chapter, we begin the transition from basic to advanced model building through the introduction of the nuanced issues and choices that a data scientist considers when building enterprise-grade models. We will discuss data splitting options, compare modeling algorithms, present a two-stage grid-search strategy for hyperparameter optimization, introduce H2O AutoML for automatically fitting multiple algorithms to data, and further investigate feature engineering tactics to extract as much information as possible from the data. We will introduce H2O Flow, a menu-based UI that is included with H2O, which is useful for monitoring the health of the H2O cluster and enables interactive data and model investigations.

Throughout the entire process, we will illustrate these advanced model-building concepts using the Lending Club problem that was introduced in *Chapter 3, Fundamental Workflow – Data to Deployable Model*. By the end of this chapter, you will be able to build an enterprise-scale, optimized predictive model using one or more supervised learning algorithms available within H2O. After that, all that is left is to review the model and deploy it into production.

In this chapter, we will cover the following main topics:

- Splitting data for validation or cross-validation and testing
- Algorithm considerations
- Model optimization with grid search
- H2O AutoML
- Feature engineering options
- Leveraging H2O Flow to enhance your IDE workflow
- Putting it all together – algorithms, feature engineering, grid search, and AutoML

Technical requirements

We are introducing the code and datasets in this chapter for the first time. At this point, if you have not set up your H2O environment, please refer to *Appendix – Alternative Methods to Launch H2O Clusters*, to do so.

Splitting data for validation or cross-validation and testing

Splitting data into training, validation, and test sets is the accepted standard for model building when the size of the data is sufficiently large. The idea behind validation is simple: most algorithms naturally overfit on training data. Here, overfitting means that some of what is being modeled are actual idiosyncrasies of that specific dataset (for instance, noise) rather than representative of the population as a whole. So, how do you correct this? Well, you can do it by creating a holdout sample, called a validation set, which is scored against during the model-building process to determine whether what is being modeled is a signal or noise. This enables things such as hyperparameter tuning, model regularization, early stopping, and more.

The test dataset is an additional holdout that is used at the end of model building to determine true model performance. Having holdout test data is critical for any model build. In fact, it is so critical that you should neither trust nor deploy a model that has not been measured against a test dataset.

An alternative to the train-validate-test split is to use a train-test split with k-fold cross-validation on the training data. Here is how that works:

1. Split the training data into k-folds, where, in our example, k is 5.

2. Fit a model with one of the folds playing the role of validation data and the other four folds being combined into training data.

3. Repeat this so that each fold is used as validation once.

    This yields five models, each validated on a different subset of the data.

The following diagram illustrates this concept nicely:

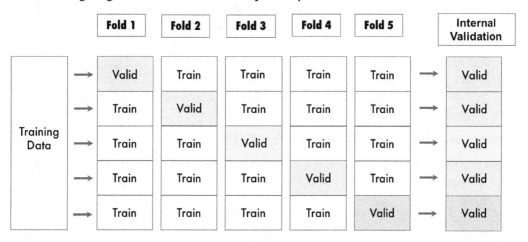

Figure 5.1 – Illustration of 5-fold cross-validation

The k-fold cross-validation approach was originally developed for small data to allow the model to see more data in training. This comes at the cost of higher computational expenses. For many data scientists, k-fold cross-validation is used regardless of the data size.

> **Model Overfitting and Data Splitting**
>
> The concept of model overfitting is critical. By definition, overfit models do not generalize well. If you are using a train-validate-test approach and building many models on the same validation set, it is likely that the leading model is overfit on the validation data. This likelihood increases as the number of models increases. Measuring the leading models against a holdout test set is the best indication of actual performance after deployment.
>
> We can minimize any overfit-to-validation issues by ensuring each model is built on its own randomly selected train-validate partition. This could occur naturally in k-fold cross-validation if each model is built on a different partitioning of data.
>
> An interesting thing happens with data science competitions that have multiple entries (in the hundreds or thousands) that are tested against a blind holdout test dataset. It has been shown that leading models commonly overfit on the test data. So, what should you do in such a situation? The obvious answer is to have an additional holdout set, such as a meta-test set, that can be used to fairly evaluate how well these models would generalize after deployment.

In the next section, we will demonstrate both approaches using the Lending Club dataset. The following code begins in the *Model training* section of *Chapter 3, Fundamental Workflow – Data to Deployable Model*, specifically in *step 3* of *Fundamental Workflow*.

Train, validate, and test set splits

We split the data into three parts: 60% for training, 20% for validation, and 20% for final testing, as shown in the following code block:

```
train, valid, test = loans.split_frame(
    seed = 25,
    ratios = [0.6, 0.2],
    destination_frames = ["train", "valid", "test"]
)
```

The preceding code is straightforward. Optionally, we set `seed` for the reproducibility of the data splits. The `ratios` parameter only requires the training and validation proportions, and the test split is obtained by subtraction from one. The `destination_frames` option allows us to name the resulting data objects, which is not required but will make their identification in H2O Flow easier.

Train and test splits for k-fold cross-validation

We could also split the data into two parts: 80% for training and 20% for testing. This can be done using a k-fold cross-validation approach, as the following code demonstrates:

```
train_cv, test_cv = loans.split_frame(
    seed = 25,
    ratios = [0.8],
    destination_frames = ["train_cv", "test_cv"]
)
```

> **How to Set a Seed**
>
> Random numbers in current computing are not random at all, but deterministic. **Pseudo-random number generators (PRNGs)** are complicated mathematical functions that return a fixed sequence of values given a specific seed. If the seed is omitted, the computer will set the seed automatically – typically, from the system clock. This seed value is often reported in logs. Setting the seed in code allows the analysis to be explicitly reproducible.

Next, we will turn our attention to choosing a modeling algorithm.

Algorithm considerations

In this section, we will address the question of how a data scientist should decide which of the many machine learning and statistical algorithms should be chosen to solve a particular problem. We assume some prior familiarity with statistical and machine learning models such as logistic regression, decision trees, random forests, and gradient boosting models.

As outlined in *Chapter 4, H2O Model Building at Scale – Capability Articulation* H2O provides multiple supervised and unsupervised learning algorithms that can be used to build models. For example, in the case of a binary classification problem, a data scientist could choose a parametric GLM model (logistic regression); semiparametric GAM; nonparametric tree-based approaches such as **Random Forest**, **GBM**, **XGBoost**, or **RuleFit**; models from the machine learning community such as **Support Vector Machines (SVMs)** or **Deep Learning Neural Networks**; or the simple **Naïve Bayes Classifier**. To complicate things even further, any subset of these algorithms could be combined into one predictive model using **Stacked Ensembles** (which is a method for combining multiple highly predictive models into a single model; we will discuss this in the *H2O AutoML* section). So, what is a data scientist to do?

A Note on RuleFit

The RuleFit algorithm is actually a penalized linear model. Here, we list it with tree-based models because the rules are extracted from a large population of randomly created decision trees. Rule selection and model regularization occur via LASSO. The intent is to combine the interpretability of linear models and explicit rules with the flexibility and predictive power of tree-based methods.

If the only criterion for model selection is pure predictive power, a data scientist could simply try everything and pick the model that performs best on a test dataset. Let's call this the *Kaggle solution*, named after the popular Kaggle data science competitions. Kaggle competitions result in algorithms and modeling approaches being pressure tested over multiple problems and datasets. Insights discovered during these competitions have found their way into real-world data science practices.

However, in an enterprise setting, it is rare for predictive power to be the only consideration for algorithm selection. Model transparency could be another. As an oversimplification, parametric models that are inherently interpretable (GLM) might be less predictive than nonparametric models. Nonparametric models such as random forest, GBM, XGBoost, and deep learning neural networks are black boxes that are difficult to interpret but frequently produce superior predictions. (Note that the GAM and RuleFit algorithms combine model transparency with predictions that often rival black-box methods.)

In addition to pure modeling criteria, there are business and implementation considerations that come into play in modeling and deployment decisions. We will cover these, in more detail, in the later chapters.

In the remaining part of this section, we will give a high-level overview of decision trees, random forest, and gradient boosting models. We will illustrate the Lending Club data while concentrating on two specific boosting implementations: H2O GBM and XGBoost.

Algorithm Popularity in the Industry

Our collective experience of working with scores of customers spanning multiple industries leads to the following general observations. First, classification problems are more prevalent than regression problems by a wide margin. Second, when choosing an algorithm, the gold standard for interpretable classification problems remains logistic regression (GLM). The most frequent nonparametric algorithm choice is some form of gradient boosting, currently the GBM, XGBoost, or LightGBM implementations. The popularity of gradient boosting has been helped by its frequent appearance (either alone or in an ensemble) high up on the Kaggle leaderboards.

An introduction to decision trees

At the heart of every random forest or GBM implementation is the concept of a decision tree. A decision tree can be used for either *classification*, where observations are assigned to discrete groups, or *regression*, where observations are a numerical outcome.

Observation assignment is made through *conditional control statements* that form a tree-like structure. The general decision tree algorithm can be described as follows:

1. Search through all the candidate predictors, identifying the variable split that yields the greatest predictive power.

2. For each newly created branch, repeat the variable splitting process from *step 1*.

3. Continue until the stopping criteria are met.

The functions used for splitting include information entropy and the Gini coefficient. Let's illustrate them using entropy. In information theory, the entropy of a random variable is the average level of uncertainty in the variable's outcomes. A pure or homogeneous classification tree node will have an entropy of zero. At each candidate split, we calculate the entropy and choose the split with the lowest entropy.

Conceptually, we could continue splitting until all nodes are pure, but that would yield an extremely overfit tree. Instead, we utilize stopping criteria such as the following:

*   The minimum number of observations that is needed at each node after splitting

*   The reduction in entropy is not enough based on a selected cutoff value

*   The maximum depth of the tree

To illustrate, let's suppose we are building a decision tree to model the probability of surviving the sinking of the Titanic in 1912. Our data includes name, gender, age, the class of passage booked, the price of the tickets, the location of the cabin or berth, the city where the passenger boarded, any traveling companions, and more. The resulting decision tree can be found in the diagram that follows.

The first split increases the predictive power the most (by reducing entropy the most):

*   Is the subject Male? If yes, the next split is created by the `Age < 18` rule.

*   For males older than 18, the survival probability for this terminal or *leaf* node is 17%.

*   For males under 18, one more split is needed: `3rd Class`.

*   For males in the 3rd class who are under 18, the survival probability is 14%.

- For males in the 1st and 2nd classes who are under 18, the survival probability is 44%.

- The tree on the `Male=Yes` branch stops splitting at these leaf nodes because one or more stopping criteria have been met.

A similar process for the `Male=No` branch proceeds. Note that according to this model, non-`3rd Class` females have a survival probability of 95%. For `3rd Class` female passengers, survival probabilities depend on where they boarded, resulting in either a 38% or 70% survival probability leaf node. The decision tree model supports the *women and children first* ethos for emergencies at sea:

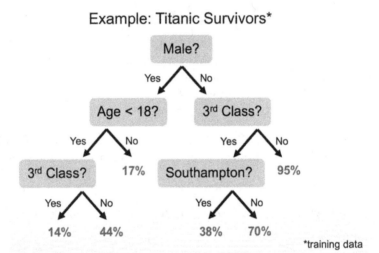

Figure 5.2 – A decision tree modeling the survival probabilities on the Titanic

Decision trees have some clear advantages. Their layout is simple to comprehend, and their interpretation, as we have just demonstrated, is straightforward. The algorithm trains and scores quickly. Decision trees are robust when it comes to nonlinear relationships, feature distributions, correlated features, and missing values. On the other hand, they do not model linear relationships efficiently. They have high variance, meaning, in part, that trees are easily overfitted. Perhaps their greatest drawback is that individual decision trees don't predict particularly well, which is an issue that was first raised by the original developers of the decision tree methodology.

To remedy the poor predictive properties of decision trees, algorithms based on ensembles of individual trees have been developed. In general, the objective of ensemble methods is to create a *strong learner* by combining information across multiple *weak learners* (in our case, decision trees). The adaptation of two ensemble methods, bagging and boosting, to trees has resulted in random forest and gradient boosting algorithms, respectively. We will review each of these ensemble methods and their implementations in H2O next.

Random forests

Bagging (which is short for *bootstrap aggregating*) is an ensemble method that fits models to bootstrapped samples of the data and averages across them. **Bootstrapping** is a resampling method that samples from the data rows with replacement. This creates randomness in the row (or observation) space. Random forest is a bagging method for decision trees that adds randomness to the column (or variable) space.

The random forest algorithm can be described as follows:

1. Build a deep tree based on randomly selected rows of the data.
2. At each split, only evaluate a random subset of variables to split on.
3. Repeat this many times, creating a *forest* as a collection of all the trees.
4. Get the average across all trees in the forest.

H2O includes two implementations of random forest, **Distributed Random Forest** (DRF) and **Extremely Randomized Trees** (XRT). In the following sections, we will summarize these algorithms.

Distributed Random Forest (DRF)

DRF is the default random forest implementation in H2O. The highlights of this algorithm are listed as follows:

- Each tree in a DRF is built in parallel.
- The splitting rule is created by choosing the most discriminative threshold among a random subset of candidate features.

Extremely Randomized Trees (XRT)

The XRT algorithm adds additional randomness to the splitting-rule process. This has the effect of reducing model variance at the cost of (slightly) increased bias. XRT is enabled by setting `histogram_type="Random"`:

- Each tree in an XRT is built in parallel.

- Rather than finding the most discriminative threshold, this algorithm will create thresholds at random for each candidate variable. The best of this set is picked as the splitting rule.

The hyperparameters for both random forest implementations are shared.

Random forest hyperparameters

The random forest methods in H2O require the following hyperparameters:

- The number of trees to be built, `ntrees` (this defaults to 50).

- The maximum tree depth, `max_depth` (this defaults to 20). Note that too large a value can result in overfitting.

- The minimum number of observations per leaf, `min_rows` (this defaults to 1).

Additional hyperparameters are available for tuning the random forest model. You can find them at `https://docs.h2o.ai/h2o/latest-stable/h2o-docs/data-science/drf.html`. A grid search can aid the process of hyperparameter selection and model optimization.

Gradient boosting

Boosting is an ensemble method that combines models sequentially, with each new model built on the residuals of the previous model. Boosted trees are based on a sequence of relatively shallow decision trees.

The boosted trees algorithm can be described as follows:

1. Start by building a shallow decision tree.
2. Fit a shallow decision tree to the residuals of the previous tree.
3. Multiply the residual tree by a shrinkage parameter (or the learning rate).
4. Repeat *steps 2* and *3* until the stopping criteria are met.

Building on the residuals makes the algorithm concentrate on areas where the model is not predicting well. The process is illustrated in the following diagram:

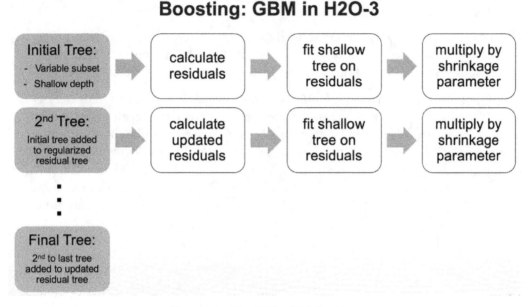

Figure 5.3 – The H2O GBM algorithm

The GBM approach results in highly predictive models, but care must be taken to avoid overfitting. H2O includes two versions of gradient boosting: H2O GBM and XGBoost. In the following sections, we will summarize these algorithms.

H2O GBM

The H2O GBM implementation follows the original algorithm, as described in the book, *The Elements of Statistical Learning by Jerome H. Friedman, Robert Tibshirani, and Trevor Hastie*, with modifications to improve performance on large and complex data. We can summarize this as follows:

- Each tree in a GBM is built in parallel.
- Categorical variables can be split into groups instead of just using Boolean splits.
- Shared histograms are used to calculate cut points.
- H2O uses a greedy search of histogram bins, optimizing the improvement in squared error.

One important advantage of this implementation is that H2O GBM naturally handles high-cardinality categorical variables (that is, categorical variables with a lot of categories).

XGBoost

XGBoost is very similar to classic GBM, with the main difference being the inclusion of a penalty term for the number of variables. Mathematically, this means it contains regularization terms in the cost function. Trees are grown in *breadth* rather than *depth*.

Another popular GBM approach is LightGBM. The LightGBM algorithm builds trees as deep as necessary by repeatedly splitting the one leaf that gives the biggest gain. Unlike XGBoost, trees are grown in *depth* rather than *breadth*. In theory, LightGBM is optimized for sparse data. While H2O does not implement LightGBM directly, it provides a method for emulating the LightGBM approach using a set of options within XGBoost (such as setting `tree_method="hist"` and `grow_policy="lossguide"`). For more details, please refer to `https://docs.h2o.ai/h2o/latest-stable/h2o-docs/data-science/xgboost.html`.

Boosting hyperparameters

All boosting methods in H2O require the following hyperparameters:

- The number of trees to be built, `ntrees` (the default is 50).
- The maximum tree depth, `max_depth` (the default is 6).
- The shrinkage parameter or learning rate, `learn_rate` (the default is 0.3).

Simply adding trees to boosting approaches without further restrictions can lead to overfitting. A grid search can aid in the process of hyperparameter tuning. Additional hyperparameters for boosting will be introduced in the *Model optimization with grid search* section.

Baseline model training

Returning to the Lending Club data, now we are ready to build baseline models for each algorithm we are considering. By baseline, we mean models that have been fitted with settings at reasonable or default values. This will be the starting point in model optimization.

As discussed in *Chapter 3*, *Fundamental Workflow – Data to Deployable Model*, we start
with the bad_loan response and the same set of predictors for all models:

```
response = "bad_loan"
omit = ["issue_d", response]
predictors = list(set(loans.columns) - set(omit))
```

In the preceding code, we remove the bad_loan response variable and the issue_d raw
date variable from the predictors. Recall that issue_d was used to create two features,
issue_d_month and issue_d_year, which are included in the predictors.

Next, we fit a baseline H2O GBM model using a train-validate-test split, followed by a
baseline XGBoost model using 5-fold cross-validation.

Baseline GBM train-validate-test model

The first model we fit is a default H2O GBM, trained on the 60%–20% training-validation
split with the following default settings:

```
from h2o.estimators.gbm import H2OGradientBoostingEstimator
gbm = H2OGradientBoostingEstimator(seed = 25)
gbm.train(x = predictors,
          y = response,
          training_frame = train,
          validation_frame = valid,
          model_id = "gbm_baseline")
```

Here, the model_id parameter in the gbm.train command is optional and used to
label the model object for identification in H2O Flow.

We will investigate model diagnostics and explainability, in greater depth, in *Chapter 7*,
Understanding ML Models. Here, we are only using a couple of those commands to aid in
comparing the gradient boosting algorithms. To begin with, we visualize the performance
of the baseline GBM model across all splits using the model_performance method:

```
%matplotlib inline
gbm.model_performance(train).plot()
```

The %matplotlib command allows figures to be displayed in a Jupyter notebook. This is only required once and is not needed outside of Jupyter. The first ROC curve is for the train split:

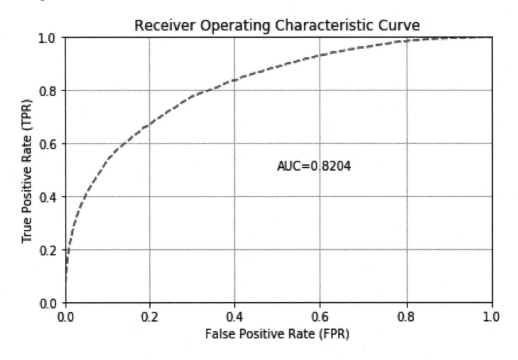

Figure 5.4 – The ROC curve for the GBM train split

The second ROC curve for the validation split uses similar code:

```
gbm.model_performance(valid).plot()
```

This will produce the following output:

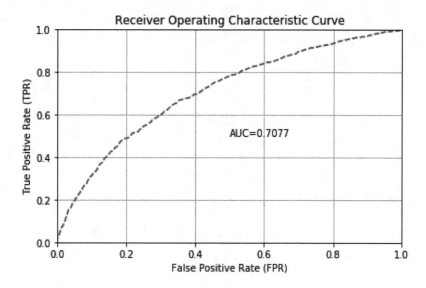

Figure 5.5 – The ROC curve for the GBM validation split

The ROC curve for the test split uses similar code:

```
gbm.model_performance(test).plot()
```

This will produce the following output:

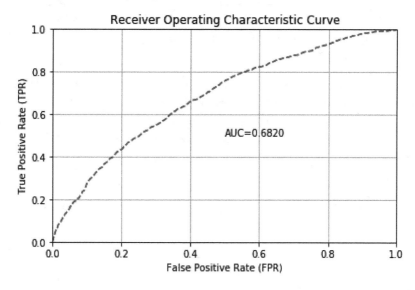

Figure 5.6 – The ROC curve for the GBM test split

To extract the AUC for these splits, we enter the following:

```
print(gbm.model_performance(train).auc(),
      gbm.model_performance(valid).auc(),
      gbm.model_performance(test).auc())
```

The code block and results, as produced in the Jupyter notebook, are shown here:

```
print(gbm.model_performance(train).auc(),
      gbm.model_performance(valid).auc(),
      gbm.model_performance(test).auc())
```

0.8204427250097287 0.707735133507756 0.6819899227739407

Figure 5.7 – The GBM model performance results from the Jupyter notebook

Additionally, the train and validation performance values are stored in the model object:

```
gbm.auc(train = True, valid = True)
```

This will return a dictionary, as follows:

```
gbm.auc(train = True, valid = True)
```

{'train': 0.820444783137194, 'valid': 0.707735133507756}

Figure 5.8 – AUC from the GBM model object in the Jupyter notebook

These results show that the baseline GBM model is overfitting on the training data. This is not a surprise.

Let's take a quick look at model interpretation, which we will cover in more depth in *Chapter 7*, *Understanding ML Models*. The variable importance plot ranks variables in terms of relative importance in predicting bad loans. Relative importance for a variable is determined by checking whether that variable was used to split on and calculating the decrease in squared error across all trees.

Here is the code to produce a variable importance plot:

```
gbm.varimp_plot(20)
```

The generated plot is as follows:

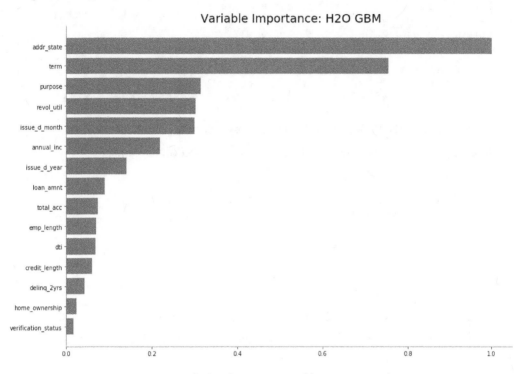

Figure 5.9 – The baseline GBM variable importance plot

The resulting variable importance plot, as shown in *Figure 5.9*, shows that the address state, which is a high-cardinality categorical variable with 50 levels corresponding to the states in the United States, is by far the most important variable.

Baseline XGBoost cross-validated model

Let's build our baseline XGBoost model using 5-fold cross-validation and the train-test split:

```
from h2o.estimators import H2OXGBoostEstimator
xgb = H2OXGBoostEstimator(nfolds = 5, seed = 25)
xgb.train(x = predictors,
          y = response,
          training_frame = train_cv,
          model_id = "xgb")
```

In the preceding code, `nfolds` sets the number of folds, `seed` is optional and included here for instructional purposes, and `model_id` is an optional identifier for use in H2O Flow.

We can get AUC for the train and cross-validation sets directly from the model object:

```
xgb.auc(train = True, xval = True)
```

This yields the following result:

```
xgb.auc(train = True, xval = True)
```

```
{'train': 0.8487580939525755, 'xval': 0.6986774438717214}
```

Figure 5.10 – The XGBoost model train and cross-validation performance results

The test set AUC requires that we include the test data to be scored against:

```
xgb.model_performance(test_cv).auc()
```

This results in the following output:

```
xgb.model_performance(test_cv).auc()
```

```
0.6834855386904981
```

Figure 5.11 – The XGBoost model test performance results from the Jupyter notebook

We can easily combine these results into a single dictionary using a little Python code:

```
perf = xgb.auc(train = True, xval = True)
perf["test"] = xgb.model_performance(test_cv).auc()
perf
```

This Python code block produces the following:

```
perf = xgb.auc(train = True, xval = True)
perf['test'] = xgb.model_performance(test_cv).auc()
perf
```

```
{'train': 0.8487580939525755,
 'xval': 0.6986774438717214,
 'test': 0.6834855386904981}
```

Figure 5.12 – XGBoost model performance as a dictionary

Again, the AUC values confirm that the baseline model is overfit on the training data and is far too optimistic. The fact that the cross-validation and test AUC values are in the same ballpark is comforting, as it means the cross-validation procedure is more accurately reflecting what you might see in the out-of-sample test data. This is an important check and might not always be the case, especially when the training and test splits cover different time periods. Next, let's consider a variable importance plot for the baseline XGBoost model:

```
xgb.varimp_plot(20)
```

The results are as follows:

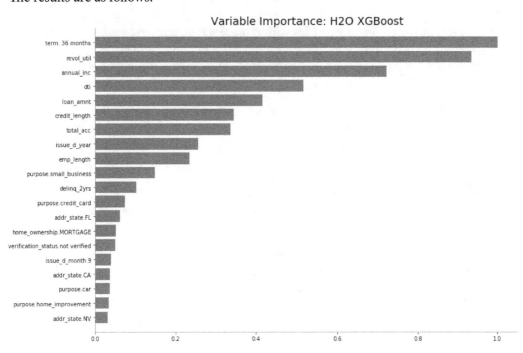

Figure 5.13 – A baseline XGBoost variable importance plot

A comparison of the variable importance plots for the GBM and XGBoost baseline models demonstrates some of the differences between these two boosting algorithms. Additionally, it introduces us to a more nuanced discussion of how to choose an algorithm given the multiple options under consideration.

Notice that the most important variable in the H2O GBM model is `addr_state`, a high-cardinality categorical variable (with approximately 50 levels corresponding to the states in the United States). XGBoost defaults to the one-hot encoding of categorical variable levels. One-hot encoding represents each level of a categorical variable with a numeric variable containing 1 for the rows in that level and 0 otherwise. The one-hot encoding of a categorical variable with 50 levels such as `addr_state` results in 50 new, relatively sparse variables corresponding to each state. In the XGBoost variable importance plot, states appear individually and far lower in importance, such as `addr_state_FL`, `addr_state_CA`, and `addr_state_NV` in the previous plot.

A data scientist could always address this issue with feature engineering approaches such as target encoding. Target encoding, which we will revisit in more detail later, is a method for replacing levels of categorical variables with representative numeric values. If target encoding is implemented, then the choice between XGBoost and H2O GBM might come down to pure performance. On the other hand, if target encoding is not an option, then H2O GBM should be the boosting algorithm choice.

In other words, XGBoost requires target encoding, while H2O GBM gives the data scientist the option of modeling the high-cardinality categorical variables directly or by using a target-encoded version of those variables. This is a nice illustration of the interaction between algorithms, feature engineering choices, and potentially other factors such as business, compliance, or regulatory considerations.

Next, we will turn our attention to improving our baseline models by using grid search to find the hyperparameter settings for model optimization.

Model optimization with grid search

Choosing an algorithm for building a predictive model is not enough. Many algorithms have hyperparameters whose values have a direct impact on the predictive power of the model. So, how do you choose values for your hyperparameters?

A brute-force method would create a grid of all possible values and search over them. This approach is computationally expensive, takes an inordinate amount of time, and ultimately, yields results that are not much better than what we could achieve by other means. We have outlined a strategy for grid search that has proved effective in building optimized models while running in a reasonable amount of time.

The general strategy entails, first, tuning a few key parameters using a **Cartesian** grid search. These key parameters are those we expect to have the biggest impact on the results. Then, we fine-tune other parameters using a random grid search. This two-stage approach allows us to hone in on the computationally expensive parameters first.

From our experience with gradient boosting methods across many datasets from many domains, our strategy follows these principles:

1. The optimal value for maximum allowed tree depth (`max_depth`) is heavily data- and problem-specific. Deeper trees, especially at depths greater than 10, can take significantly longer to train. In the interest of time, it is a good idea to, first, narrow the approximate depth down to a small range of values.

2. We increase the number of trees (`ntrees`) until the validation set error starts increasing.

3. Very low learning rates (`learn_rate`) are universally recommended. This generally yields better accuracy but requires more trees and additional computation time. A clever alternative is to start with a relatively high learning rate (say 0.05 or 0.02) and iteratively shrink it by using `learn_rate_annealing`. For example, setting `learn_rate=0.02` and `learn_rate_annealing=0.995` speeds up convergence significantly without sacrificing much accuracy. This is very useful for hyperparameter searches. For even faster scans, values of 0.05 and 0.99 can be tried instead.

4. Sampling rows and columns using `sample_rate` and `col_sample_rate`, respectively, reduces the validation and test set error rates and improves generalization. A good starting point for most datasets is between 70% and 80% sampling on both rows and columns (rates of between 0.7 and 0.8). Optionally, the column sampling rate per tree parameter (`col_sample_rate_per_tree`) can be set. It is multiplicative with `col_sample_rate`. For example, setting both parameters to 0.9 results in an overall 81% of columns being considered for the split.

5. Early stopping using `stopping_rounds`, `stopping_metric`, and `stopping_tolerance` can make grid search more efficient. For our needs, we can use 5, AUC, and 1e-4 as good starting points. This means that if the validation AUC doesn't improve by more than 0.0001 after 5 iterations, the computation will end.

6. To improve the predictive accuracy of highly imbalanced classification datasets, the `sample_rate_per_class` parameter can be set. This implements stratified row sampling based on the specific response class. The parameter values are entered as an array of ratios, one per response class, in lexicographic order.

7. Most of the other options only have a relatively small impact on model performance. That said, they might be worth tuning with a random hyperparameter search.

Next, we will build an optimized H2O GBM model for the Lending Club data and compare the results with the baseline model.

Step 1 – a Cartesian grid search to focus on the best tree depth

The optimal max_depth parameter value is very specific to the use case and data being modeled. Additionally, it has a profound impact on the model training time. In other words, large tree-depth values require significantly more computation than smaller values. First, we will focus on good candidate max_depth values using a quick Cartesian grid search.

Here, we use early stopping in conjunction with learning rate annealing to speed up convergence and efficiently tune the max_depth parameter:

1. We start by defining the hyperparameters:

    ```
    from h2o.grid.grid_search import H2OGridSearch
    hyperparams = {
        "max_depth": list(range(2, 14, 2))
    }
    ```

2. We follow our strategy by defining an excessive number of trees with early stopping enabled. We use learning rate annealing, as shown in the following code block, to shrink the learn_rate and sample 80% of the rows and columns. Also, we score every 10 trees in order to make the early stopping reproducible. For model builds with large amounts of data, we might want to score every 100 or 1,000 trees:

    ```
    gbm_grid = H2OGradientBoostingEstimator(
        ntrees = 10000,
        stopping_metric = "AUC",
        stopping_rounds = 5,
        stopping_tolerance = 1e-4,
        learn_rate = 0.05,
        learn_rate_annealing = 0.99,
        sample_rate = 0.8,
        col_sample_rate = 0.8,
        score_tree_interval = 10,
        seed = 25
    )
    ```

> **Setting the score_tree_interval Parameter**
>
> Scoring trees during a model grid search is essentially a waste of compute resources, as it requires more time to arrive at an optimum solution. However, it is required to make the early stopping process reproducible. We want to set the value high enough to ensure reproducibility but also not waste compute cycles. This is, largely, data- and problem-specific. The value of 10 that we used earlier is perhaps too aggressive even for this problem; 100 might have been more appropriate.

3. Now we define the grid and set the search criteria to Cartesian:

```
grid = H2OGridSearch(
    gbm_grid,
    hyperparams,
    grid_id = "gbm_depth_grid",
    search_criteria = {"strategy": "Cartesian"}
)
```

4. Then, we fit the grid, as shown in the following code block:

```
grid.train(x = predictors,
           y = response,
           training_frame = train,
           validation_frame = valid)
```

5. To display the grid search results based on descending values of AUC, we use the following code:

```
sorted_grid = grid.get_grid(
    sort_by = "auc", decreasing = True)
print(sorted_grid)
```

This results in the following:

```
sorted = grid.get_grid(sort_by = 'auc', decreasing = True)
print(sorted)

  max_depth                  model_ids                 auc
0         4  gbm_depth_grid_model_2  0.7111529277230403
1         2  gbm_depth_grid_model_1  0.7077429040968353
2         6  gbm_depth_grid_model_3   0.706764773666835
3         8  gbm_depth_grid_model_4  0.6983977768163934
4        10  gbm_depth_grid_model_5  0.6910955916183174
5        12  gbm_depth_grid_model_6  0.6844227036614173
```

Figure 5.14 – Tuning the maximum tree depth parameter value

For this data and the H2O GBM algorithm, the `max_depth` values of 2 to 6 appear to give the best results. Next, we will search in the range of 2 to 6 and tune any additional parameters.

Step 2 – a random grid search to tune other parameters

Now that we have focused on a good range for the maximum tree depth, we can set up our tuning hyperparameters as follows:

```
hyperparams_tune = {
    "max_depth" : list(range(2, 6, 1)),
    "sample_rate" : [x/100. for x in range(20,101)],
    "col_sample_rate" : [x/100. for x in range(20,101)],
    "min_split_improvement": [0, 1e-8, 1e-6, 1e-4]
}
```

The `min_split_improvement` parameter attempts to reduce overfitting in the GBM and XGBoost models by demanding that each split does not lead to worse error measures. We will try four different settings of that parameter.

In the following search criteria, we limit our runtime to 5 minutes for illustrative purposes. Additionally, we limit the number of models built to 10. Depending on your use case, you might want to increase the runtime substantially or just exclude these options altogether:

```
search_criteria_tune = {
    "strategy" : "RandomDiscrete",
    "max_runtime_secs" : 300,
    "max_models" : 10,
    "stopping_rounds" : 5,
    "stopping_metric" : "AUC",
    "stopping_tolerance" : 1e-3
}
```

Also, we set up our final grid parameters:

```
gbm_final_grid = H2OGradientBoostingEstimator(
    ntrees = 10000,
    learn_rate = 0.05,
    learn_rate_annealing = 0.99,
    score_tree_interval = 10,
    seed = 12345
)
```

And we fit our final grid, as shown in the following code block:

```
final_grid = H2OGridSearch(
    gbm_final_grid,
    hyper_params = hyperparams_tune,
    grid_id = "gbm_final_grid",
    search_criteria = search_criteria_tune)
```

> **Further Documentation**
>
> There are several additional hyperparameters available that are listed in the H2O documentation at http://docs.h2o.ai/h2o/latest-stable/h2o-docs/parameters.html.
>
> Further details on grid search can be found at http://docs.h2o.ai/h2o/latest-stable/h2o-docs/grid-search.html#grid-search-in-python.

Now we train the model. Note that the max_runtime_secs setting, as follows, overrides the value set in search_criteria_tune:

```
final_grid.train(
    x = predictors,
    y = response,
    max_runtime_secs = 180,
    training_frame = train,
    validation_frame = valid
)
```

After 3 minutes or less, we look at the results of our grid search sorted by AUC:

```
grid = final_grid.get_grid(sort_by = "auc",
                           decreasing = True)
grid
```

The output is as follows:

	col_sample_rate	max_depth	min_split_improvement	sample_rate	\
0	0.44	4	0.0	0.84	
1	0.26	4	1.0E-6	0.91	
2	0.45	3	1.0E-8	0.69	
3	0.86	4	1.0E-8	0.92	
4	0.67	5	0.0	0.4	
5	0.61	4	1.0E-6	0.48	
6	0.71	4	1.0E-6	0.3	
7	0.72	2	1.0E-8	0.68	
8	0.41	2	0.0	0.77	
9	0.46	2	1.0E-4	0.48	

	model_ids	auc
0	gbm_final_grid_model_8	0.7131296692054205
1	gbm_final_grid_model_3	0.7124726628869076
2	gbm_final_grid_model_9	0.7117921640897826
3	gbm_final_grid_model_4	0.7116096456021055
4	gbm_final_grid_model_1	0.7115625401706325
5	gbm_final_grid_model_10	0.7107034382989326
6	gbm_final_grid_model_7	0.708596946050065
7	gbm_final_grid_model_5	0.7082204037839034
8	gbm_final_grid_model_6	0.7080174046738467
9	gbm_final_grid_model_2	0.7065488838121041

Figure 5.15 – The grid search results for GBM model optimization

Optimization Strategy Results

This exercise shows the importance of hyperparameter tuning. Although we constrained this optimization by only searching for 3 minutes and producing 10 models, 9 out of 10 outperformed the baseline GBM model with default values.

We can easily select the best model based on the previous leaderboard and extract its AUC performance values:

```
best_gbm = grid.models[0]
perf = best_gbm.auc(train = True, valid = True)
perf["test"] = best_gbm.model_performance(test).auc()
perf
```

This Python code block produces the following:

```
best_gbm = grid.models[0]
perf = best_gbm.auc(train = True, valid = True)
perf['test'] = best_gbm.model_performance(test).auc()
perf
```

```
{'train': 0.7715007027429149,
 'valid': 0.7131296692054205,
 'test': 0.6917964044086415}
```

Figure 5.16 – The performance of the best optimized GBM model from the grid search

Our grid search strategy is a tremendous way to fine-tune the hyperparameters of a machine learning model. Next, we will explore AutoML in H2O.

H2O AutoML

The most efficient method of model building and tuning utilizes H2O AutoML. AutoML builds models from multiple algorithms while implementing appropriate grid search and model optimization based on the model type. The user can specify constraints such as compute time limits or limits on the number of models created.

Some features of AutoML include the following:

- AutoML trains a random grid of GLMs, GBMs, and DNNs using a carefully chosen hyperparameter space.

- Individual models are tuned using a validation set or with cross-validation.

- Two stacked ensemble models are trained by default: *All Models* and a lightweight *Best of Family* ensemble.

- AutoML returns a sorted leaderboard of all models.

- Any model can be easily promoted to production.

Stacked ensembles are highly predictive models that usually appear at the top of leaderboards. Similar to the other ensemble approaches that we introduced earlier (such as bagging and boosting), we stack works by combining information from multiple predictive models into one. Unlike bagging and boosting, which rely on weak learners as the component models, stacking works by optimally combining a diverse set of strongly predictive models. The *All Models* stacked ensemble is created by combining the entire list of models investigated in an AutoML run. The *Best of Family* ensemble contains, at most, six component models. Its performance is usually comparable to the All Models ensemble, but being less complex, it is typically better suited for production. (For more information on stacked ensembles, see `https://docs.h2o.ai/h2o/latest-stable/h2o-docs/data-science/stacked-ensembles.html`).

Training models using AutoML is relatively straightforward:

```
from h2o.automl import H2OAutoML
aml = H2OAutoML(max_models = 10,
                max_runtime_secs_per_model = 60,
                exclude_algos = ["DeepLearning"],
                seed = 25)
aml.train(x = predictors,
          y = response,
          training_frame = train_cv)
```

> **The AutoML Runtime Parameter Choices**
>
> Our values for the `max_runtime_secs_per_model` and `max_models` parameters allow us to quickly screen multiple model types while restricting overall runtime. This is neither optimal nor recommended and is used in tutorial or classroom settings to demonstrate AutoML. Instead, you can set the overall `max_runtime_secs` parameter to an explicit value. The default is 3,600 (that is, 1 hour).

H2O AutoML trains the following algorithms (in order):

- Three XGBoost GBMs
- A grid of GLMs
- A DRF
- Five H2O GBMs
- A deep neural net

- An extremely randomized forest
- Random grids of XGBoost GBMs, H2O GBMs, and deep neural nets
- Two stacked ensemble models

If there is not enough time to complete all of these algorithms, some can be omitted from the leaderboard.

The AutoML leaderboard

The AutoML object contains a leaderboard of models along with their cross-validated model performance. You can create a leaderboard for a specific dataset by specifying the `leaderboard_frame` argument.

The models are ranked by a metric whose default is based on the problem type:

- For regression, this is deviance.
- For binary classification, AUC is the default metric.
- For multiclass classification, we use the mean per-class error.
- Additional metrics such as Logloss are provided for convenience.

Next, we print out the leaderboard:

```
print(aml.leaderboard)
```

model_id	auc	logloss	aucpr	mean_per_class_error	rmse	mse
StackedEnsemble_AllModels_AutoML_20210707_153116	0.713355	0.400888	0.333119	0.35009	0.351147	0.123304
StackedEnsemble_BestOfFamily_AutoML_20210707_153116	0.712794	0.401124	0.332364	0.351302	0.351211	0.123349
GBM_1_AutoML_20210707_153116	0.703838	0.404872	0.322928	0.351571	0.352786	0.124458
XGBoost_3_AutoML_20210707_153116	0.701885	0.406119	0.319911	0.353353	0.353532	0.124985
GLM_1_AutoML_20210707_153116	0.699046	0.406926	0.315193	0.358155	0.353472	0.124942
GBM_5_AutoML_20210707_153116	0.698736	0.407134	0.320094	0.359607	0.353676	0.125087
GBM_2_AutoML_20210707_153116	0.696458	0.407998	0.316976	0.363644	0.354166	0.125433
GBM_3_AutoML_20210707_153116	0.692507	0.41101	0.308223	0.364154	0.355506	0.126384
DRF_1_AutoML_20210707_153116	0.690973	0.414376	0.30815	0.363189	0.355066	0.126072
GBM_4_AutoML_20210707_153116	0.683994	0.416615	0.302158	0.36352	0.357366	0.12771

Figure 5.17 – The AutoML leaderboard

As expected, the stacked ensemble models outperform all the individual models on the leaderboard. Any of these models can be selected for further investigation and potential deployment. Next, we will show you how to select the top model.

Examining the top model

The `aml.leader` object contains the best model from the leaderboard, including details for both training and cross-validated data. We use the following code to print the AUC values for training, cross-validation, and testing data for the best model:

```
best = aml.leader
perf = best.auc(train = True, xval = True)
perf["test"] = best.model_performance(test_cv).auc()
perf
```

The resulting values are as follows:

```
best = aml.leader
perf = best.auc(train = True, xval = True)
perf['test'] = best.model_performance(test_cv).auc()
perf

{'train': 0.8982787353045297,
 'xval': 0.7133552099266027,
 'test': 0.6979120005522923}
```

Figure 5.18 – The performance of the best model from the AutoML leaderboard

Examining a selected model

In practice, the leading model might not be the one you end up putting into production. As alluded to earlier, other considerations such as the modeling type, regulatory or compliance requirements, internal business preferences, and the likelihood of model approval could play a role in determining which model to use.

> **Other Reasons to Use a Leaderboard**
>
> The most obvious reason for using AutoML and exploring its leaderboard is to find the top model and put that into production. As mentioned earlier, that might not be allowed. Let's consider a scenario where I am only able to put a GLM into production. So, why bother fitting other models using AutoML? One answer is that the best model gives me a practical ceiling that I can also report. *GLM has an AUC of 0.69905, while the best possible model yielded 0.71336.*
>
> In a business context, I should always be able to translate performance differences into terms of cost reduction or increased profit. In other words, the AUC difference translated into dollars and cents is "the cost of doing business" or "how much money is being left on the table" by using the selected model instead of the best.

Here, we demonstrate how to select any model from the leaderboard. The top individual (non-ensemble) model is in third position. We select this model with the following code and examine its performance in terms of AUC:

```
select = h2o.get_model(aml.leaderboard[2, "model_id"])
perf = select.auc(train = True, xval = True)
perf["test"] = select.model_performance(test_cv).auc()
perf
```

This results in the following:

```
select = h2o.get_model(aml.leaderboard[2, 'model_id'])
perf = select.auc(train = True, xval = True)
perf['test'] = select.model_performance(test_data=test).auc()
perf
```

```
{'train': 0.821048177023111,
 'xval': 0.7038378194723536,
 'test': 0.6890479203139213}
```

Figure 5.19 – The performance of the selected model from the AutoML leaderboard

Once a model object has been selected via AutoML, all the model diagnostics and explainability procedures, which we will cover in *Chapter 7, Understanding ML Models,* will be available.

Feature engineering options

In this section, we will demonstrate how feature engineering can lead to better predictive models. Second only to data cleaning, typically, feature engineering is the most time-consuming of all tasks involved in the modeling process. It can also be the "secret sauce" that makes for a great predictive model.

So, what does feature engineering mean? Put simply, it is how to extract information from raw data into a form that is both usable by the modeling algorithm and interpretable for the problem at hand. For example, a date or date-time object might be represented in data as a string or a number (for example, Unix time is the number of seconds since 00:00:00 UTC on January 1, 1970). Presented with such features, an algorithm is liable to treat dates as levels of a categorical variable or a continuous numeric value. Neither of these forms is very helpful. However, embedded in this raw data is information about not only the day, the month, and the year, but the day of the week, the weekend or weekday, seasons, holidays, and more. If the object contains time, then you could also produce the hour of the day, the time of day (for example, morning, afternoon, evening, or night), and more.

Which features to create depends largely on the use case. Even in the best of circumstances, most engineered features may not be selected by a model algorithm. Subject-matter expertise and an understanding of the context of the problem play a major role in engineering good features. For example, the debt-to-income ratio used in lending divides how much a customer owes per month by their monthly income. This engineered feature has proven so predictive in risk modeling that it has been given its own name and abbreviation, DTI.

Subject-Matter Expertise and Feature Engineering

One of our colleagues, an accomplished data scientist, multiple Kaggle grandmaster, and a Ph.D., once commented that he did not enjoy FinTech data science competitions because "there is more Fin than Tech in them." By this, he meant, at least in part, that those problems put a premium on subject-matter insights that he had no experience of.

Another great example of feature engineering is **natural language processing** (**NLP**) in the context of predictive modeling. NLP attempts to represent words, word meanings, and sentences as numeric values that can be incorporated naturally into machine learning algorithms. TF-IDF and word embeddings (word2vec) are two such approaches. We will cover these in more detail in the *Modeling in Sparkling Water* section of *Chapter 6, Advanced Model Building – Part II*, and in the detailed Lending Club analysis within *Chapter 8, Putting It All Together*.

In the remainder of this section, we will investigate target encoding in depth. Target encoding is one of the most common and impactful feature engineering options available. We will illustrate its use in the **Lending Club model**. In *Chapter 8, Putting It All Together*, we will implement additional feature engineering recipes to improve the predictive model.

Target encoding

Target encoding replaces categorical levels with a numeric value representing some function of the target variable, such as the mean. The following diagram illustrates mean target encoding:

A	0.75 (3 out of 4)
B	0.66 (2 out of 3)
C	1.00 (2 out of 2)

Feature	Outcome	MeanEncode
A	1	0.75
A	0	0.75
A	1	0.75
A	1	0.75
B	1	0.66
B	1	0.66
B	0	0.66
C	1	1.00
C	1	1.00

Figure 5.20 – Mean target encoding

The approach is simple: replace the categorical feature levels (**A**, **B**, and **C**) with their respective means (**0.75**, **0.66**, and **1.00**).

Target encoding is a clever idea and, in spirit, is analogous to the random effects found in statistical random and mixed effect models. In fact, for certain simple cases, you can prove that mean target encoding actually yields the empirical Bayes estimates of the random effects. What this means is that the intent behind target encoding is based on sound principles.

However, target encoding uses a function of the target as an input to predict the target. This is the very definition of data leakage. Data leakage leads to overly optimistic models that do not generalize well and are, at best, misleading in practice. H2O implements target encoding using carefully constructed cross-validation procedures. Essentially, this eliminates data leakage by calculating the target-encoded value for each row based on other folds of the data.

> **Random Effects**
>
> The mathematical structure underlying the estimation of random effects in statistical models does not suffer from data leakage concerns in the same way that target encoding does. This is because the information in the target variable is partitioned, and the portion used to estimate the random effects is separate from the portion used to estimate the other model parameters.

We use the **H2O-3 Target Encoding Estimator** to replace categorical values with a mean of the target variable. We tune target encoding via the following:

- Setting data_leakage_handling to k-fold controls data leakage.
- Adding random noise to the target average helps to prevent overfitting.
- We adjust for categories with small group sizes through blending.

Any categorical levels with fewer observations will result in an unreliable (high variance) target-encoded mean. A blended average consisting of a weighted average of the group's target value and the global target value can improve this estimate. By setting blending=True, the target mean will be weighted based on the sample size of the categorical level.

When blending is enabled, the smoothing parameter controls the rate of transition between the level's posterior probability and the prior probability (with a default value of 20). The inflection_point parameter represents half of the sample size for which we completely trust the estimate. The default value is 10.

Target encoding the Lending Club data

To determine whether a categorical variable would benefit from target encoding, first, create a table for the variable, which has been sorted from most frequent to least frequent. To do this efficiently, we will define a Python function:

```python
import numpy as npdef sorted_table(colname, data = train_cv):
    tbl = data[colname].table().as_data_frame()
    tbl["Percent"] = np.round((100 * tbl["Count"]/data.nrows),
2)
    tbl = tbl.sort_values(by = "Count", ascending = 0)
    tbl = tbl.reset_index(drop = True)
    return(tbl)
```

Note that the preceding code requires the Python pandas package to be available since the as_data_frame call outputs the table in a pandas format.

First, consider the purpose variable, which records the purpose of the loan:

```python
sorted_table("purpose")
```

This returns the following:

	purpose	Count	Percent
0	debt_consolidation	14484	46.30
1	credit_card	4117	13.16
2	other	3346	10.69
3	home_improvement	2287	7.31
4	major_purchase	1718	5.49
5	small_business	1448	4.63
6	car	1118	3.57
7	wedding	732	2.34
8	medical	539	1.72
9	moving	475	1.52
10	educational	331	1.06
11	house	325	1.04
12	vacation	290	0.93
13	renewable_energy	75	0.24

`sorted_table("purpose")`

Figure 5.21 – Levels of the purpose variable

Note the high concentration of loans for debt consolidation (46%) and the sizable numbers for both credit cards (13%) and other (11%), with the remaining 30% captured among 11 other loan purposes. One option for this data would be to collapse the categories into fewer levels and leave the purpose variable as a categorical variable. This might make sense if the categories could be collapsed in a coherent manner. A better option uses mean target encoding to represent all levels without overfitting those with small percentages in the tail. Blending will also be enabled here, although the amount of smoothing it provides might not be impactful. The renewable_energy category has 75 observations, which, in most cases, is sufficient to reliably estimate a mean even though the percentage is very small.

A second variable to consider is addr_state:

```
sorted_table("addr_state")
```

The first few rows are listed as follows:

	addr_state	Count	Percent
0	CA	5559	17.77
1	NY	2956	9.45
2	FL	2285	7.30
3	TX	2179	6.96
4	NJ	1444	4.62
5	IL	1227	3.92
6	PA	1190	3.80
7	VA	1097	3.51
8	GA	1097	3.51
9	MA	1042	3.33
10	OH	962	3.07

Figure 5.22 – The top ten states by count

And the last few rows are listed as follows:

43	TN	24	0.08
44	MS	21	0.07
45	IN	14	0.04
46	NE	9	0.03
47	ID	9	0.03
48	IA	8	0.03
49	ME	3	0.01

Figure 5.23 – The last seven states by count

High-cardinality categorical variables such as `addr_state` are prime candidates for target encoding. The distribution of records is also highly skewed, with the top 4 levels accounting for, approximately, 40% of the data. Blending will be especially important because the raw counts for states in the tail are extremely small:

1. Start by importing the target encoding estimator and specifying the columns to encode:

    ```
    from h2o.estimators import H2OTargetEncoderEstimator
    encoded_columns = ["purpose", "addr_state"]
    ```

2. The `k_fold` strategy requires a fold column, which is created as follows:

    ```
    train_cv["fold"] = train_cv.kfold_column(
        n_folds=5, seed=25)
    ```

3. Train a target encoding model by setting the parameters:

    ```
    te = H2OTargetEncoderEstimator(
        data_leakage_handling = "k_fold",
        fold_column = "fold",
        noise = 0.05,
        blending = True,
        inflection_point = 10,
        smoothing = 20
    )
    ```

    Here is the training:

    ```
    te.train = (x = encoded_columns, y = response,
        training_frame = train_cv)
    ```

4. Now, create a new target-encoded train and test set, explicitly setting the noise level on the test set to 0:

    ```
    train_te = te.transform(frame = train_cv)
    test_te = te.transform(frame = test_cv, noise = 0.0)
    ```

5. Next, check the results of target encoding by looking at histograms of the target-encoded variables:

    ```
    train_te["purpose_te"].hist()
    ```

This yields the following plot:

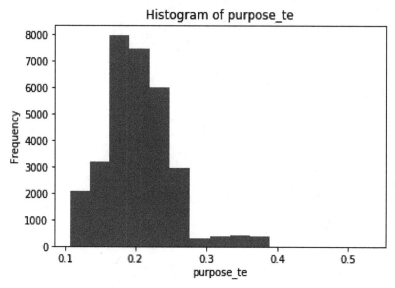

Figure 5.24 – The target-encoded loan purpose variable

The following code produces a histogram for the `addr_state_te` variable:

```
train_te["addr_state_te"].hist()
```

The output is as follows:

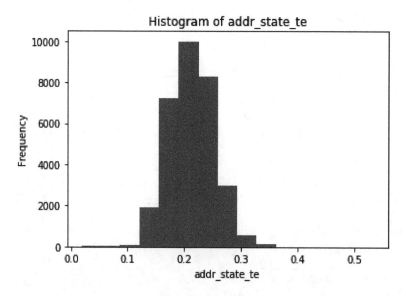

Figure 5.25 – The target-encoded address state variable

6. Add the target-encoded variables to the predictor list:

```
predictors.extend(["addr_state_te", "purpose_te"])
```

7. Then, remove the source columns, using a list comprehension for efficiency:

```
drop = ["addr_state", "purpose"]
predictors = [x for x in predictors if x not in drop]
```

8. As we create other features, our predictor list will change. In order to keep track of these steps, it is wise to update a copy of the `predictors` list rather than the original:

```
transformed = predictors.copy()
```

9. Additionally, we rename our datasets using the target-encoded values for convenience:

```
train = train_te
test = test_te
```

How Much Should You Tune the Target Encoding Model?

Note that we used the same target encoding parameters for transforming two different variables. So, why not encode variables individually with custom parameter settings for each? In our situation, we did not need to. The only parameter values to vary are those that determine the amount of blending: `inflection_point` and `smoothing`. For the `purpose` variable, blending is not really needed because sample sizes are large enough to yield accurate means. On the other hand, the `addr_state` variable would greatly benefit from blending. Therefore, we set the parameters to values that are reasonable for `addr_state`. These will, essentially, be ignored by `purpose`.

In situations where the outputs of one model are inputs for another, always bear in mind that what matters is the effect that varying parameter settings in the input model have on the final model's predictions.

---

10. Let's refit our model with these new features using AutoML and print the leaderboard:

```
check = H2OAutoML(max_models = 10,
                  max_runtime_secs_per_model = 60,
                  exclude_algos = ["DeepLearning"],
                  seed = 25)
check.train(x = transformed,
```

```
                y = response,
                training_frame = train)
        check.leaderboard
```

This results in the following output:

model_id	auc	logloss	aucpr	mean_per_class_error	rmse	mse
StackedEnsemble_AllModels_AutoML_20210709_131836	0.71158	0.402041	0.325923	0.347285	0.351822	0.123779
StackedEnsemble_BestOfFamily_AutoML_20210709_131836	0.709824	0.402592	0.324423	0.348619	0.352037	0.12393
GBM_1_AutoML_20210709_131836	0.704491	0.405415	0.316446	0.35398	0.353315	0.124831
GBM_2_AutoML_20210709_131836	0.702125	0.405913	0.318244	0.349798	0.35341	0.124899
GBM_5_AutoML_20210709_131836	0.702095	0.40581	0.318142	0.359754	0.353281	0.124808
XGBoost_3_AutoML_20210709_131836	0.70016	0.407282	0.313654	0.353429	0.354187	0.125449
GBM_3_AutoML_20210709_131836	0.698733	0.407999	0.310791	0.356246	0.354498	0.125669
GBM_4_AutoML_20210709_131836	0.691322	0.412464	0.299487	0.361668	0.356547	0.127125
GLM_1_AutoML_20210709_131836	0.690441	0.409719	0.304832	0.360625	0.354623	0.125757
DRF_1_AutoML_20210709_131836	0.690031	0.412513	0.302618	0.361108	0.355708	0.126528

Figure 5.26 – The AutoML leaderboard after target encoding

The best individual (non-ensemble) model is a GBM, whose performance (**0.704491**) is only slightly better than the best GBM (**0.703838**) prior to target encoding. People often ask, is this tiny performance gain worth the effort of target encoding? That question misses the point entirely. Recall that H2O GBM naturally handles high-cardinality categorical variables well, so the fact that performance is equivalent should come as no surprise.

11. What is the right question to ask? Let's look at variable importance and compare the variables before and after target encoding:

```
check_gbm = h2o.get_model(check.leaderboard[2, "model_
id"])
check_gbm.varimp_plot(15)
```

Before target encoding, the high-cardinality variables are among the most important:

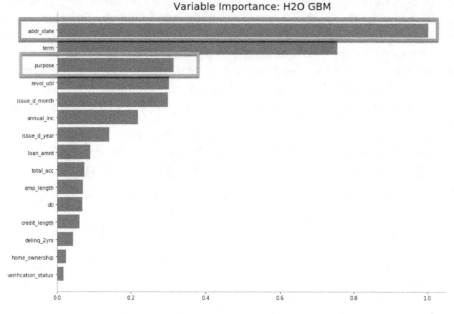

Figure 5.27 – Variable importance for the H2O GBM model before target encoding

After target encoding, the importance of these categorical variables has changed:

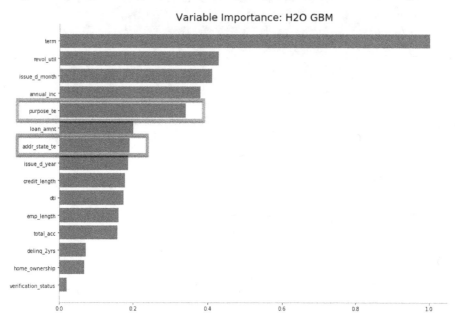

Figure 5.28 – Variable importance for the H2O GBM model after target encoding

The effect of target encoding for the GBM model has less to do with overall model performance than with the impact and interpretation of those variables. Target encoding `purpose` has only slightly changed its importance, from *third* place to *fifth* place. Target encoding `addr_state` has decreased its impact substantially, from *first* place to *seventh* place. This impact difference also leads to an interpretability difference. The former model primarily splits on state, in essence implying a different loan default model per state (with implications that might have to be explained to a regulator). In the latter model, the effect of the state is adjusted in a very similar manner to random effects in a statistical model.

The data scientist has the option of choosing which scenario makes the most sense for their situation. An additional benefit of target encoding `addr_state` is the blending feature, which, in production, will generalize better for states with low counts.

Select the best XGBoost target-encoded model from the leaderboard:

```
check_xgb = h2o.get_model(check.leaderboard[5, "model_id"])
check_xgb.varimp_plot(15)
```

This yields the following plot:

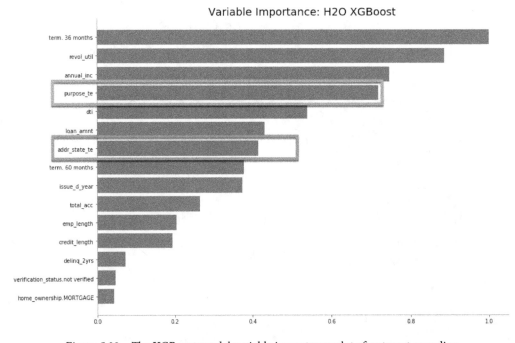

Figure 5.29 – The XGBoost model variable importance plot after target encoding

Both `purpose` and `address_state` have entered the top 10 in positions almost identical to the GBM model. Target encoding categorical variables is more important in XGBoost models than in GBM models. Other considerations being equal, some feature engineering steps may be influenced by the algorithm chosen.

Other feature engineering options

There are multiple ways in which to categorize feature engineering options, and there are almost an infinite number of approaches you could take, depending on the problem. For some high-level categorizations, we can think of the following rough hierarchy:

- Algebraic transformers:

  1. Add, subtract, multiply, or divide numeric columns to create new interaction features.

  2. Use simple mathematical functions such as log, exp, power, roots, and trigonometric functions

- Cluster-based transformers: Use k-means or other unsupervised algorithms to create clusters. Then, do the following:

  1. Measure the distance of a numeric observation to a specified cluster.

  2. Consider each cluster as a level of a categorical variable and target encode to clusters.

- Numeric to categorical transformations: Often, binning into deciles or using histograms and then taking the mean within each bin produces good predictive features.

- Categorical to numeric transformations:

  1. One-hot or indicator value encoding.

  2. Target encoding.

  3. Numeric summary encoding: This is similar to target encoding except you are summarizing one of the numeric predictor columns rather than the target variable; for example, the mean temperature per state.

4. **Weight of evidence**: This is only used for binary classification problems. The weight of evidence is the natural log of the ratio of successes over failures (good over bad and ones over zeros):

$$WOE = ln\left(\frac{1's}{0's}\right)$$

- Dimension reduction transformations: Truncated eigenvalue or singular value decomposition.

As a data scientist, you can combine multiples of these components into a reasonable feature for a particular problem at hand. We will revisit some of these recipes in our complete analysis of the Lending Club data, which can be found in *Chapter 8, Putting It All Together*.

Leveraging H2O Flow to enhance your IDE workflow

H2O Flow is a web-based UI available wherever an H2O cluster is running. Flow is interactive, allowing users to do everything including import data, build models, investigate models, and put models into production. While incredibly easy to use, our experience is that most data scientists (authors included) prefer coding in Python or R to menu-driven interactive interfaces. This section is written for those data scientists: why use Flow when I am a coder?

There are two main reasons:

- **Monitoring** the state of the H2O cluster and the jobs that are being run
- **Interactive investigation** of the data, models, model diagnostics, and more where interactivity is an asset rather than an annoyance

Connecting to Flow

By default, Flow is started on port 54321 of the H2O server as the cluster is launched (this port is configurable at startup). Enter `Error! Hyperlink reference not valid.` into your browser to open Flow. The Flow UI is straightforward and self-explanatory, with helpful instructions and videos:

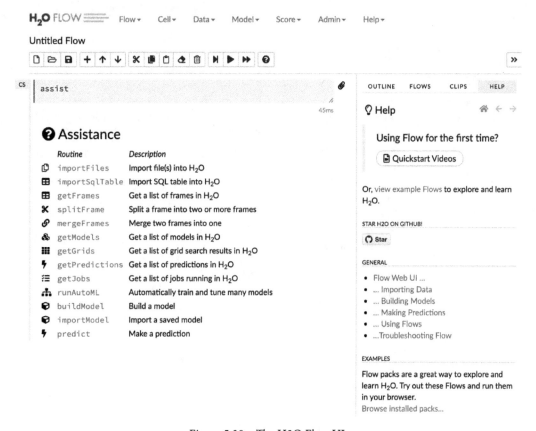

Figure 5.30 – The H2O Flow UI

First, let's consider Flow's monitoring capabilities.

Monitoring with Flow

Under the **Admin** menu in Flow, the top three options are **Jobs**, **Cluster Status**, and **Water Meter**. These are central to the monitoring capabilities of Flow, and we will review each of them individually.

The Flow **Admin** menu is shown here:

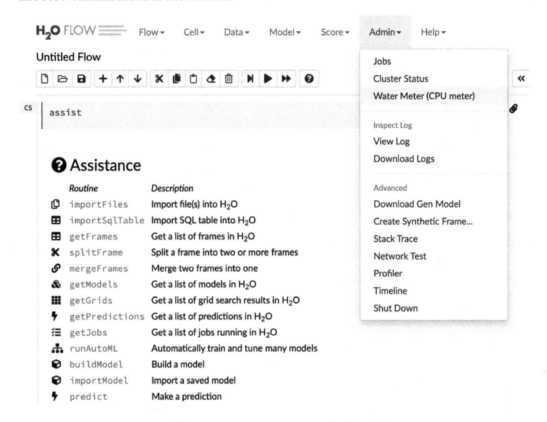

Figure 5.31 – The monitoring options using the Flow Admin menu

We start by monitoring jobs.

Monitoring jobs

The **Jobs** option lists jobs for data frames, models, grid search, AutoML, and more as they launch, while they are running, and after completion. Clicking on the **Refresh** button will instruct the UI to continually update the jobs, which is especially useful when H2O is executing a grid search or AutoML run. Rather than using the menu, you can also enter getJobs in the Flow command line:

Figure 5.32 – Listing the job options using the Flow Admin menu

We continue by monitoring health.

Monitoring H2O cluster health

The **Cluster Status** option is available from the drop-down menu or the Flow command line using the `getCloud` command. This monitors the health of the cluster and is one of the first places to check whether H2O does not appear to be working correctly:

Figure 5.33 – Monitoring the cluster status using the Flow Admin menu

Next, we will monitor CPU usage.

Monitoring CPU usage live

The **Water Meter** tool is a useful monitor of CPU usage. It shows a bar per CPU with colors corresponding to the activity status of each CPU. Rather than watching a black progress bar grow across the cell of a Jupyter notebook, the Water Meter is much more informative. Also, it illustrates, in real time, how well a particular computation is distributed among the available compute resources:

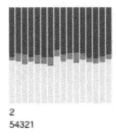

Legend

Each bar represents one CPU.

Blue: idle time
Green: user time
Red: system time
White: other time (e.g. i/o)

Figure 5.34 – The H2O Flow Water Meter

We can also monitor grid search.

Monitoring grid search

H2O Flow allows you to interactively monitor individual model builds, but it is especially useful when executing multiple jobs like a grid search or AutoML creates. These can be monitored live upon launch and reviewed during runtime and after completion.

The first step in our grid search strategy was to evaluate model depth. While the model is running, we can open Flow and list jobs. The job named `gbm_depth_grid` is running. Clicking on the name opens the running job, allowing us to view more details or cancel the job. These actions are not readily available from within Python:

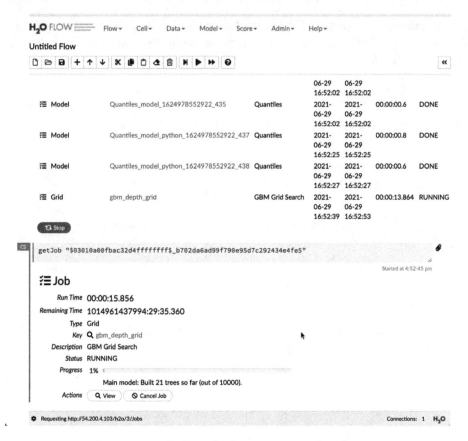

Figure 5.35 – Grid search job monitoring within Flow

Selecting the **View** button at any time opens the grid:

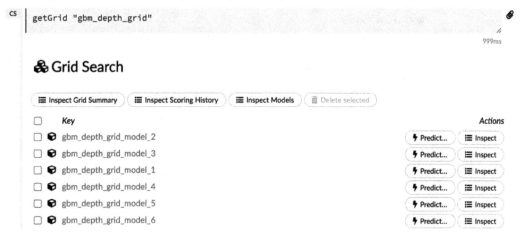

Figure 5.36 – Grid search results within Flow

The subsequent selection of any of the individual grid models opens an interactive model view, which we will discuss in more detail in the next section and in *Chapter 7, Understanding ML Models*.

Monitoring AutoML

Monitoring AutoML jobs is similar. First, search for the AutoML build job in the job listings and select the model's name link:

				16:57:11	16:57:11
≔ Auto Model	AutoML_20210629_215711705@@bad_loan		AutoML build	2021-06-29	2021-06-29
				16:57:11	16:58:12
≔ Model	XGBoost_1_AutoML_20210629_215711		XGBoost def_1	2021-06-29	2021-06-29
				16:57:11	16:57:18
≔ Model	Quantiles_model_1624978552922_4032		Quantiles	2021-06-29	2021-06-29

Figure 5.37 – Selecting the AutoML build job

Once the AutoML build is in process, you can monitor the progress live or click on **View** to watch the leaderboard as the models are built:

Figure 5.38 – Viewing the AutoML build job

> **Note: Flow is Great for Monitoring Leaderboards**
>
> The interactive leaderboard is a great way to monitor AutoML jobs in real time. This is especially true for AutoML runs that are not constrained to finish quickly but plan to run for multiple hours as models are built. Again, all that is available in Python is a progress bar that can seem very slow if you cannot see the actual work on the server (Figure 5.39).

≔ Leaderboard

⟳ Monitor Live

▾ M O D E L S

models sorted in order of auc, best first

	model_id	auc	logloss	aucpr
0	StackedEnsemble_AllModels_AutoML_20210709_050356	0.7133552099266027	0.4008875123992031	0.3331190515037253
1	StackedEnsemble_BestOfFamily_AutoML_20210709_050356	0.7127942536163684	0.40112416561020175	0.3323636598463481
2	GBM_1_AutoML_20210709_050356	0.7038378194723536	0.4048718730109695	0.3229283085231189
3	XGBoost_3_AutoML_20210709_050356	0.7018852717798446	0.4061193478699023	0.319911345567044
4	GLM_1_AutoML_20210709_050356	0.6990461687138205	0.4069256926189401	0.31519328501171445
5	GBM_5_AutoML_20210709_050356	0.6987364437762574	0.4071340312777311	0.32009378522975007
6	GBM_2_AutoML_20210709_050356	0.696457526175039	0.40799819243916674	0.3169759160602985
7	GBM_3_AutoML_20210709_050356	0.6925073212178198	0.411009590593443	0.30822319188695096
8	DRF_1_AutoML_20210709_050356	0.6909730327435186	0.4143762844098374	0.3081500977501284
9	GBM_4_AutoML_20210709_050356	0.6839942327763355	0.4166145383620577	0.3021583701931105
10	XGBoost_1_AutoML_20210709_050356	0.6750682772559911	0.42741956090525046	0.2848358536542451
11	XGBoost_2_AutoML_20210709_050356	0.6722874901581177	0.4332617130333989	0.2833423182534677

Figure 5.39 – The AutoML leaderboard in Flow

The selection of any individual AutoML model opens an interactive model view.

Interactive investigations with Flow

As we mentioned earlier, interactivity in Flow is quite useful for the live monitoring of running jobs. In addition, Flow makes exploring data before modeling and evaluating candidate models after model build more convenient than coding in Python. The only potential downside to menu-driven exploration is when reproducibility is at a premium and documentation of the whole modeling process is required. We will explore this topic in more detail when we discuss H2O AutoDoc capabilities in *Chapter 7, Understanding ML Models*.

Interactive data exploration in Flow

Perform the following steps:

1. In the **Data** menu, select **List All Frames**:

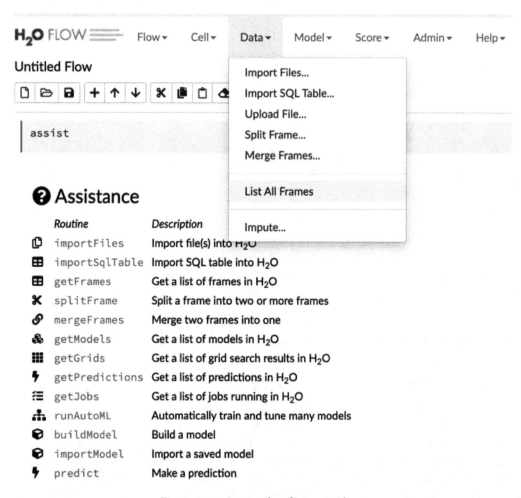

Figure 5.40 – Listing data frames in Flow

2. Click on the **LendingClubClean.hex** link to pull up the data summary:

```
getFrameSummary "LendingClubClean.hex"
```
158ms

▦ LendingClubClean.hex

Actions: [▦ View Data] [✕ Split] [⊘ Build Model] [⊹ Run AutoML] [⚡ Predict] [⬇ Download] [🔒 Export]   [🗑 Delete]

Rows	Columns	Compressed Size
38980	16	1MB

▾ COLUMN SUMMARIES

label	type	Missing	Zeros	+Inf	-Inf	min	max	mean	sigma	cardinality	Actions
loan_amnt	int	1	0	0	0	500.0	35000.0	10699.4196	7184.9438	·	Convert to enum
term	enum	1	31528	0	0	0	1.0	0.1912	0.3932	2	Convert to numeric
int_rate	real	1	0	0	0	5.4200	24.4000	11.9449	3.6464	·	·
emp_length	int	1007	4761	0	0	0	10.0	4.8481	3.5581	·	Convert to enum
home_ownership	enum	1	17056	0	0	0	4.0	·	·	5	Convert to numeric
annual_inc	real	5	0	0	0	1896.0	6000000.0	68668.9202	64802.9177	·	·
verification_status	enum	1	17947	0	0	0	1.0	0.5396	0.4984	2	Convert to numeric
issue_d	int	1	0	0	0	1180656000000.0	1322697600000.0	1283943368644.6548	32388482262.2405	·	Convert to enum
purpose	enum	1	1411	0	0	0	13.0	·	·	14	Convert to numeric
addr_state	enum	1	77	0	0	0	49.0	·	·	50	Convert to numeric
dti	real	1	193	0	0	0	29.9900	13.2682	6.7309	·	·
delinq_2yrs	int	30	34597	0	0	0	13.0	0.1532	0.5149	·	Convert to enum
revol_util	real	91	994	0	0	0	119.0	48.8202	28.4275	·	·
total_acc	int	30	0	0	0	1.0	90.0	22.0238	11.6372	·	Convert to enum
bad_loan	int	0	32776	0	0	0	1.0	0.1592	0.3658	·	Convert to enum
credit_length	int	30	11	0	0	0	65.0	13.5037	6.8634	·	Convert to enum

[← Previous 20 Columns] [→ Next 20 Columns]

Figure 5.41 – The Lending Club data in Flow

Clicking on the `purpose` column link produces a summary plot:

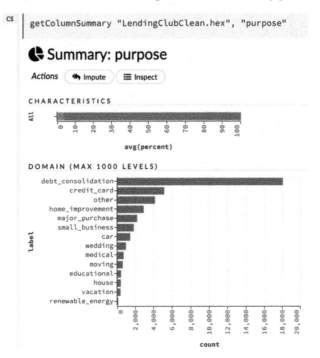

Figure 5.42 – The loan purpose data column in Flow

3. Next, clicking on `Inspect` and then `domain` will yield a summary table similar to the one that we created in Python:

```
inspect getColumnSummary "LendingClubClean.hex", "purpose"
```

ⓘ Data

TABLES

Name	Actions	
ⓘ characteristics	☰ Inspect	⅃⊥ Plot
ⓘ domain	☰ Inspect	⅃⊥ Plot

```
grid inspect "domain", getColumnSummary "LendingClubClean.hex", "purpose"
```

label	count	percent
debt_consolidation	18026	46.24422780913289
credit_card	5119	13.132375577219086
other	4111	10.546434068753207
home_improvement	2872	7.367880964597229
major_purchase	2154	5.525910723447922
small_business	1788	4.586967675731144
car	1411	3.6198050282196
wedding	938	2.4063622370446383
medical	686	1.7598768599281682
moving	593	1.5212929707542329
educational	417	1.0697793740379682
house	393	1.008209338122114
vacation	374	0.9594663930220626
renewable_energy	97	0.24884556182657774

Figure 5.43 – A table of the loan purpose data in Flow

Model exploration in Flow

Selecting any model in Flow, whether through the **List All Models** option in the **Model** menu item, from a grid search, or the AutoML leaderboard, yields a model summary:

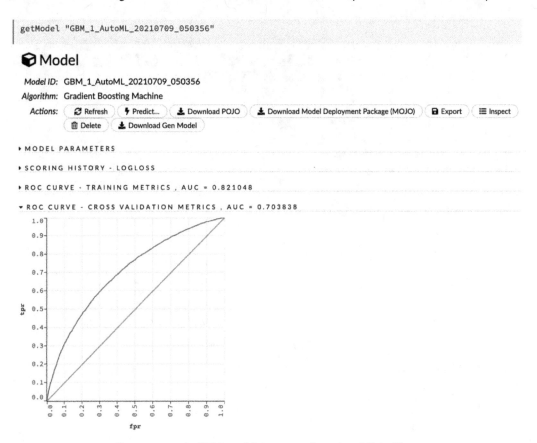

Figure 5.44 – A GBM model summary from AutoML in Flow

The layout of the model summary makes it very easy to explore. ROC curves and AUC values for training and validation sets are displayed by default. Variable importance plots are also readily available:

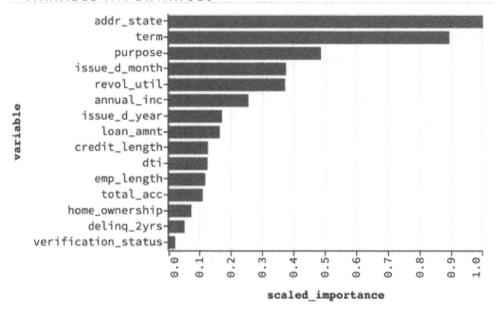

Figure 5.45 – GBM variable importance in Flow

Generally, getting results immediately through the model summary is more convenient than doing the equivalent from the Python client.

Best Practices in Flow

If you are coding in Python, we strongly suggest using Flow solely as a monitoring platform and read-only tool. That is the approach we use in our own work. Code should contain all the steps that import data, create features, fit models, deploy models, and more. This allows you to repeat any analysis and is a prerequisite for reproducibility. Code is often less convenient for investigative and interactive steps. Reserve those for Flow.

Putting it all together – algorithms, feature engineering, grid search, and AutoML

The H2O AutoML implementation is simple yet powerful, so why would we ever need grid search? In fact, for a lot of real-world enterprise use cases, any of the top candidates in an AutoML leaderboard would be great models to put into production. This is especially true of the stacked ensemble models produced by AutoML.

However, our coverage of grid search was not just to satisfy intellectual curiosity. A more involved process, which we will outline next, uses AutoML followed by a customized grid search to discover and fine-tune model performance.

An enhanced AutoML procedure

Here are the steps:

1. Start by running AutoML on your data to create a baseline leaderboard. You can investigate leading models, gain an understanding of the runtimes required to fit algorithms to your data, and more, which may inform future AutoML parameter choices and expectations.

2. The second stage is feature engineering. While developing new features, repeat AutoML runs as desired to check the impact of engineering and see what other insights might be gained from diagnostics.

3. After completion of the feature engineering stage, use AutoML to create a final leaderboard.

4. Choose a model from the leaderboard as a candidate for production. If you select an ensemble model, you are done. There is very little you can do to improve upon the performance of a stacked ensemble.

5. If you choose an individual model, say a GBM or DRF, use the parameters of that model as a guide for further grid searching, employing the general strategy outlined in this chapter. It is possible to further fine-tune a candidate model using additional grid search.

This enhanced AutoML procedure might be overkill for a lot of problems. If you are in a business that has a practice of quickly building and deploying models, especially one that updates or replaces models frequently, then this approach might literally be more effort than it is worth. The advantages of a model built on recent data often outweigh the gains made by using these extra modeling steps.

However, if you are in an industry where the model review and due diligence process is long and involved, where the models that are put into production tend to stay in production for a long time, or you are working on a model that is high risk in any way (for example, one that directly impacts peoples' lives rather than just what ad they will see on a website), then this more involved procedure might well be worth the extra effort. We have used it successfully in multiple real-world use cases.

Summary

In this chapter, we have considered different options for splitting data, explored, in some depth, powerful and popular algorithms such as gradient boosting and random forest, learned how to optimize model hyperparameters using a two-stage grid search strategy, utilized AutoML to efficiently fit multiple models, and further investigated options for feature engineering, including a deep dive into target encoding. Additionally, we saw how Flow can be used to monitor the H2O system and investigate data and models interactively. You now have most of the tools required to build effective enterprise-scale predictive models using the H2O platform.

However, we are not finished with our advanced modeling topics. In *Chapter 6, Advanced Model Building – Part II*, we will discuss best practices for data acquisition, look in more depth at checkpointing and refitting models, and show you how to ensure reproducibility. Additionally, we will thoroughly consider two more hands-on examples: the first demonstrating Sparkling Water pipelines for efficiently integrating Spark capabilities with H2O modeling, and the second introducing isolation forests, an unsupervised learning algorithm for anomaly detection in H2O.

6
Advanced Model Building – Part II

In the previous chapter, *Chapter 5, Advanced Model Building – Part I*, we detailed the process for building an enterprise-grade **supervised learning** model on the H2O platform. In this chapter, we round out our advanced model-building topics by doing the following:

- Demonstrating how to build H2O supervised learning models within an Apache Spark pipeline

- Introducing H2O's **unsupervised learning** method

- Discussing best practices for updating H2O models

- Documenting requirements to ensure H2O model reproducibility

We begin this chapter by introducing Sparkling Water pipelines, a method for embedding H2O models natively within a Spark pipeline. In enterprise settings where Spark is heavily utilized, we have found this to be a popular method for building and deploying H2O models. We demonstrate by building a Sparkling Water pipeline for **sentiment analysis** using data from online reviews of Amazon food products.

We then introduce the unsupervised learning methods available in H2O. Using credit card transaction data, we build an anomaly detection model using isolation forests. In this context, the unsupervised model would be used to flag suspicious credit card transactions in a financial fraud-prevention effort.

We conclude this chapter by addressing issues pertinent to models built in this chapter, as well as in *Chapter 5, Advanced Model Building – Part I.* These are best practices for updating H2O models and ensuring reproducibility of H2O model results.

The following topics will be covered in this chapter:

- Modeling in Sparkling Water
- UL methods in H2O
- Best practices for updating H2O models
- Ensuring H2O model reproducibility

Technical requirements

The code and datasets we introduce in this chapter can be found in the GitHub repository at `https://github.com/PacktPublishing/Machine-Learning-at-Scale-with-H2O`. If you have not set up your H2O environment at this point, see *Appendix – Alternative Methods to Launch H2O Clusters for This Book,* to do so.

Modeling in Sparkling Water

We saw in *Chapter 2, Platform Components and Key Concepts,* that Sparkling Water is simply H2O-3 in an Apache Spark environment. From the Python coder's point of view, H2O-3 code is virtually identical to Sparkling Water code. If the code is the same, why have a separate section for modeling in Sparkling Water? There are two important reasons, as outlined here:

- Sparkling Water enables data scientists to leverage Spark's extensive data processing capabilities.
- Sparkling Water provides access to production Spark pipelines. We expand upon these reasons next.

Spark is rightly known for its data operations that effortlessly scale with increasing data volume. Since the presence of Spark in an enterprise setting is now almost a given, data scientists should add Spark to their skills toolbelt. This is not nearly as hard as it seems, since Spark can be operated from Python (using PySpark) with data operations written primarily in Spark SQL. For experienced Python and **Structured Query Language** (**SQL**) coders, this is a very easy transition indeed.

In the Lending Club example from *Chapter 5*, *Advanced Model Building – Part I*, data munging and **feature engineering** tasks were carried out using native H2O commands on the H2O cluster. These H2O data commands work in Sparkling Water as well. However, in an enterprise that has invested in a Spark data infrastructure, replacing H2O data commands with their Spark equivalents makes a lot of sense. It would then pass the cleaned dataset to H2O to handle the subsequent modeling steps. This is our recommended workflow in Sparkling Water.

In addition, Spark pipelines are frequently used in enterprise production settings for **extract, transform, and load** (**ETL**) and other data processing tasks. Sparkling Water's integration of H2O algorithms into Spark pipelines allows for seamless training and deployment of H2O models in a Spark environment. In the remainder of this section, we show how Spark pipelines can be combined with H2O modeling to create a Sparkling Water pipeline. This pipeline is easily promoted into production, a topic that we return to in detail in *Chapter 10*, *H2O Model Deployment Patterns*.

Introducing Sparkling Water pipelines

Figure 6.1 illustrates the Sparkling Water pipeline training and deployment process. The pipeline starts with an input data source for model training. Data cleaning and feature engineering steps are built sequentially from Spark transformers, with the outputs of one transformer becoming the inputs of the subsequent transformer. Once the dataset is in a modeling-ready format, H2O takes over to specify and build a model. We wrap all the transformer and model steps into a pipeline that is trained and then exported for production.

In the production environment, we import the pipeline and introduce new data to it (in the following diagram, we assume this happens via a data stream, but the data could be arriving in batches as well). The pipeline outputs H2O model predictions:

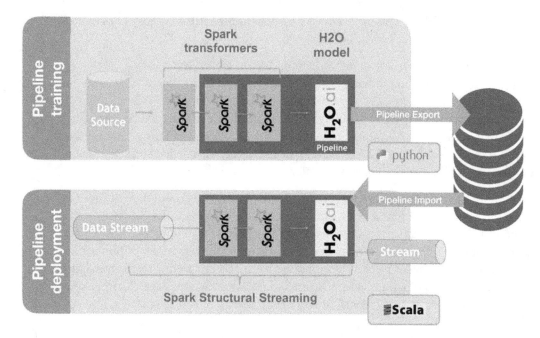

Figure 6.1 – Sparkling Water pipeline train and deploy illustration

Next, let's create a pipeline to implement sentiment analysis.

Implementing a sentiment analysis pipeline

We next create a Sparkling Water pipeline for an sentiment analysis classification problem. Sentiment analysis is used to model whether a customer has positive or negative feelings toward a product or company. It typically requires **natural language processing** (**NLP**) to create predictors from text. For our example, we use a preprocessed version of the *Amazon Fine Food reviews* dataset from the **Stanford Network Analysis Platform** (**SNAP**) repository. (See https://snap.stanford.edu/data/web-FineFoods.html for the original data.)

Let's first verify that Spark is available on our system.

```
spark
```

The following screenshot shows the output:

SparkSession - in-memory
SparkContext

Spark UI
Version
`v3.0.0`
Master
`local[*]`
AppName
`pyspark-shell`

Figure 6.2 – Spark startup in Jupyter notebook with PySparkling kernel

You can see in the Spark output that the **SparkSession** has been started and the **SparkContext** initiated.

PySpark and PySparkling

PySpark is Apache's Python interface for Spark. It provides a shell for interactive Spark sessions and access to Spark components such as Spark SQL, DataFrames, and Streaming. **PySparkling** is the H2O extension of **PySpark**, enabling H2O services to be started on a Spark cluster from Python. Our Jupyter notebook uses a PySpark shell.

In the *internal backend* mode of Sparkling Water, H2O resources piggyback on their Spark counterparts, all within the same **Java virtual machine (JVM)**. As illustrated in the following diagram, an **H2OContext** that sits on top of the **SparkContext** is launched and H2O is initialized in each worker node of the Spark cluster:

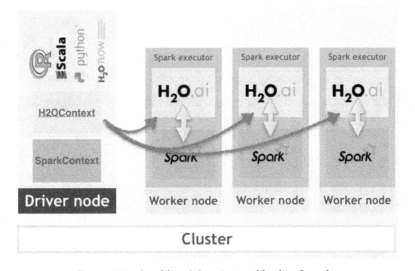

Figure 6.3 – Sparkling Water internal backend mode

PySparkling is used to create an H2OContext and initialize worker nodes, as follows:

```
from pysparkling import *
hc = H2OContext.getOrCreate()
```

This results in the following output:

Connecting to H2O server at http://30e34279d3ed:54323/h2o ... successful.

H2O_cluster_uptime:	09 secs
H2O_cluster_timezone:	Etc/GMT
H2O_data_parsing_timezone:	UTC
H2O_cluster_version:	3.32.1.3
H2O_cluster_version_age:	21 days, 10 hours and 14 minutes
H2O_cluster_name:	sparkling-water-h2o_local-1623298375712
H2O_cluster_total_nodes:	1
H2O_cluster_free_memory:	7.010 Gb
H2O_cluster_total_cores:	16
H2O_cluster_allowed_cores:	16
H2O_cluster_status:	locked, healthy
H2O_connection_url:	http://30e34279d3ed:54323/h2o
H2O_connection_proxy:	null
H2O_internal_security:	False
H2O_API_Extensions:	XGBoost, Algos, Amazon S3, Sparkling Water REST API Extensions, AutoML, Core V3, TargetEncoder, Core V4
Python_version:	3.6.13 final

Figure 6.4 – Sparkling Water cluster immediately after launch

After the H2O server is launched, we interact with it using Python commands. We will start by importing the raw data.

Importing the raw Amazon data

We import the Amazon training data into a `reviews_spark` Spark DataFrame, as follows:

```
datafile = "AmazonReviews_Train.csv"
reviews_spark = spark.read.load(datafile, format="csv",
    sep=",", inferSchema="true", header="true")
```

As an alternative, we could have imported the data using H2O and then converted the `reviews_h2o` H2O frame to the `reviews_spark` Spark DataFrame, like so:

```
import h2o
reviews_h2o = h2o.upload_file(datafile)
reviews_spark = hc.as_spark_frame(reviews_h2o)
```

This approach has the advantage of allowing us to use H2O Flow for interactive data exploration, as demonstrated in *Chapter 5, Advanced Model Building – Part I*, before converting to a Spark DataFrame.

Next, we print out the data schema to show the input variables and variable types. This is done with the following code:

```
reviews_spark.printSchema()
```

The resulting data schema is shown in the following screenshot:

```
root
 |-- Id: integer (nullable = true)
 |-- ProductId: string (nullable = true)
 |-- UserId: string (nullable = true)
 |-- ProfileName: string (nullable = true)
 |-- HelpfulnessNumerator: string (nullable = true)
 |-- HelpfulnessDenominator: string (nullable = true)
 |-- Score: string (nullable = true)
 |-- Time: string (nullable = true)
 |-- Summary: string (nullable = true)
 |-- Text: string (nullable = true)
```

Figure 6.5 – Schema for Amazon Fine Food raw data

For simplicity, we use only the `Time`, `Summary`, and overall `Score` columns in this analysis. `Time` is a date-time string, `Score` is an integer value between 1 and 5, from which sentiment is derived, and `Summary` is a short text summary of the product review. Note that the `Text` column contains the actual product review. A better model choice would include `Text` in place of—or perhaps in addition to—`Summary`.

Save the input data schema into the `schema.json` file using the following code:

```
with open('schema.json','w') as f:
    f.write(str(reviews_spark.schema.json()))
```

Saving the input data schema will make the deployment of the Sparkling Water pipeline a very simple matter.

> **Input Data and Production Data Structure**
>
> Saving the data schema for deployment presumes that production data will use the same schema. As a data scientist building a Sparkling Water pipeline, we strongly recommend that your training input data exactly follows the production data schema. It is worth the extra effort to track this information down prior to model build rather than having to reengineer something at the deployment stage.

Defining Spark pipeline stages

Spark pipelines are created by stringing individual data operations or transformers together. Each transformer takes as its input the output data from the preceding stage, which makes development for a data scientist very simple. A large job can be broken down into individual tasks that are daisy-chained together.

Apache Spark operates through lazy evaluation. This means that computations do not execute immediately; rather, operations are cached and execution occurs when an action of some sort is triggered. This approach has many advantages, including allowing Spark to optimize its compute.

In our example, all the data cleaning and feature engineering steps will be created through Spark transformers. The pipeline is finalized by training an H2O XGBoost model. For clarity, we will define a stage number for each transformer as we proceed in building the pipeline.

Stage 1 – Creating a transformer to select required columns

The Spark `SQLTransformer` class allows us to use SQL to munge data. The fact that most data scientists are already experienced with SQL makes for smooth adoption of Spark for data operations. `SQLTransformer` will be widely used in this pipeline. Run the following code to import the class:

```
from pyspark.ml.feature import SQLTransformer
```

Define a `colSelect` transformer, like so:

```
colSelect = SQLTransformer(
    statement="""
    SELECT Score,
            from_unixtime(Time) as Time,
            Summary
    FROM __THIS__""")
```

In the preceding code, we select the Score, Time, and Summary columns, converting the timestamp to a readable date-time string. __THIS__ in the FROM statement references the output from the previous transformer stage. Since this is the first stage, __THIS__ refers to the input data.

During development, it is helpful to check results at each stage by calling the transformer directly. This makes it easy to debug transformer code and understand which inputs will be available for the next stage. Calling the transformer will cause Spark to execute it along with all unevaluated upstream code. The following code snippet illustrates how to call the transformer:

```
selected = colSelect.transform(reviews_spark)
selected.show(n=10, truncate=False)
```

The first few rows are shown in the following screenshot:

```
+-----+-------------------+----------------------------------------------+
|Score|Time               |Summary                                       |
+-----+-------------------+----------------------------------------------+
|5    |2011-04-27 00:00:00|Good Quality Dog Food                         |
|1    |2012-09-07 00:00:00|Not as Advertised                             |
|4    |2008-08-18 00:00:00|"""Delight"" says it all"                     |
|2    |2011-06-13 00:00:00|Cough Medicine                                |
|5    |2012-10-21 00:00:00|Great taffy                                   |
|4    |2012-07-12 00:00:00|Nice Taffy                                    |
|5    |2012-06-20 00:00:00|Great!  Just as good as the expensive brands!|
|5    |2012-05-03 00:00:00|Wonderful, tasty taffy                        |
|5    |2011-11-23 00:00:00|Yay Barley                                    |
|5    |2012-10-26 00:00:00|Healthy Dog Food                              |
+-----+-------------------+----------------------------------------------+
only showing top 10 rows
```

Figure 6.6 – Results from the colSelect stage 1 transformer

This first transformer has taken the original data and boiled it down to three columns. We will operate on each column separately to create our modeling-ready dataset. Let's begin with the Time column.

Stage 2 – Defining a transformer to create multiple time features

The goal of this model is to predict sentiment: was the review positive or negative? Date and time are factors that could arguably influence sentiment. Perhaps people give better reviews on Friday evenings because there is a weekend upcoming.

The `Time` column is stored internally as a timestamp. To be useful in modeling, we need to extract the date-and-time information in a format that is understandable by the predictive algorithms we employ. We define an `expandTime` transformer using SparkSQL data methods (such as `hour`, `month`, and `year`) to engineer multiple new features from the raw timestamp information, as follows:

```
expandTime = SQLTransformer(
    statement="""
    SELECT Score,
           Summary,
           dayofmonth(Time) as Day,
           month(Time) as Month,
           year(Time) as Year,
           weekofyear(Time) as WeekNum,
           date_format(Time, 'EEE') as Weekday,
           hour(Time) as HourOfDay,
           IF(date_format(Time, 'EEE')='Sat' OR
               date_format(Time, 'EEE')='Sun', 1, 0) as
               Weekend,
        CASE
           WHEN month(TIME)=12 OR month(Time)<=2 THEN 'Winter'
           WHEN month(TIME)>=3 OR month(Time)<=5 THEN 'Spring'
           WHEN month(TIME)>=6 AND month(Time)<=9 THEN 'Summer'
           ELSE 'Fall'
        END as Season
    FROM __THIS__""")
```

Note that `Score` and `Summary` are selected in the `expandTime` code, but we do not operate on them. This simply passes those columns along to subsequent transformers. We engineer several features from the `Time` column: `Day`, `Month`, `Year`, `WeekNum`, `Weekday`, `HourOfDay`, `Weekend`, and `Season`. And once more, `__THIS__` refers to the output from the `colSelect` stage 1 transformer.

To check our progress in development and perhaps debug our code, we inspect the output of the second stage, which uses as its input the first-stage results stored in `selected`, as illustrated in the following code snippet:

```
expanded = expandTime.transform(selected)
expanded.show(n=10)
```

The output is shown in the following screenshot:

```
+-----+--------------------+---+-----+----+-------+-------+---------+-------+------+
|Score|             Summary|Day|Month|Year|WeekNum|Weekday|HourOfDay|Weekend|Season|
+-----+--------------------+---+-----+----+-------+-------+---------+-------+------+
|    5|Good Quality Dog ...| 27|    4|2011|     17|    Wed|        0|      0|Spring|
|    1|      Not as Advertised|  7|    9|2012|     36|    Fri|        0|      0|Spring|
|    4|"""Delight"" says...| 18|    8|2008|     34|    Mon|        0|      0|Spring|
|    2|       Cough Medicine| 13|    6|2011|     24|    Mon|        0|      0|Spring|
|    5|         Great taffy| 21|   10|2012|     42|    Sun|        0|      1|Spring|
|    4|          Nice Taffy| 12|    7|2012|     28|    Thu|        0|      0|Spring|
|    5|Great!  Just as g...| 20|    6|2012|     25|    Wed|        0|      0|Spring|
|    5|Wonderful, tasty ...|  3|    5|2012|     18|    Thu|        0|      0|Spring|
|    5|         Yay Barley| 23|   11|2011|     47|    Wed|        0|      0|Spring|
|    5|    Healthy Dog Food| 26|   10|2012|     43|    Fri|        0|      0|Spring|
+-----+--------------------+---+-----+----+-------+-------+---------+-------+------+
only showing top 10 rows
```

Figure 6.7 – Results from the expandTime stage 2 transformer

The output confirms that we have successfully replaced the Time column with a collection of newly created features.

Stage 3 – Creating a response from Score while removing neutral reviews

In this stage, we create our Sentiment response variable using values from the Score column. We could model *positive* versus *not positive* as the response, but we choose to remove neutral reviews (Score=3) and compare Positive with Negative. This is a standard approach in **net promoter score** (**NPS**) analyses and is common in sentiment analysis. It makes sense because we assume that records with a neutral response contain little information the model could learn from.

We create our createResponse transformer like so:

```
createResponse = SQLTransformer(
    statement="""
    SELECT IF(Score < 3,'Negative', 'Positive') as Sentiment,
           Day, Month, Year, WeekNum, Weekday, HourOfDay,
           Weekend, Season, Summary
    FROM  __THIS__  WHERE Score != 3""")
```

The `IF` statement assigns scores of 1 or 2 to `Negative` sentiment and all others to `Positive`, filtering out neutral reviews with the `WHERE` clause. Now, inspect the results of this intermediate step by running the following code:

```
created = createResponse.transform(expanded)
created.show(n=10)
```

This results in the following output:

```
+---------+---+-----+----+-------+-------+---------+-------+------+--------------------+
|Sentiment|Day|Month|Year|WeekNum|Weekday|HourOfDay|Weekend|Season|             Summary|
+---------+---+-----+----+-------+-------+---------+-------+------+--------------------+
| Positive| 27|    4|2011|     17|    Wed|        0|      0|Spring|Good Quality Dog ...|
| Negative|  7|    9|2012|     36|    Fri|        0|      0|Spring|     Not as Advertised|
| Positive| 18|    8|2008|     34|    Mon|        0|      0|Spring|"""Delight"" says...|
| Negative| 13|    6|2011|     24|    Mon|        0|      0|Spring|       Cough Medicine|
| Positive| 21| 10|2012|     42|    Sun|        0|      1|Spring|          Great taffy|
| Positive| 12|    7|2012|     28|    Thu|        0|      0|Spring|           Nice Taffy|
| Positive| 20|    6|2012|     25|    Wed|        0|      0|Spring|Great!  Just as g...|
| Positive|  3|    5|2012|     18|    Thu|        0|      0|Spring|Wonderful, tasty ...|
| Positive| 23| 11|2011|     47|    Wed|        0|      0|Spring|          Yay Barley|
| Positive| 26| 10|2012|     43|    Fri|        0|      0|Spring|     Healthy Dog Food|
+---------+---+-----+----+-------+-------+---------+-------+------+--------------------+
only showing top 10 rows
```

Figure 6.8 – Results from the createResponse stage 3 transformer

The only remaining feature engineering steps are those replacing the text in the `Summary` column with appropriately representative numerical values. Stages 4 through 8 will leverage Spark's built-in NLP data transformation capabilities to create features based on the text in `Summary`. While this is not a formal deep dive into NLP, we will describe each transformation step in enough detail to make our model understandable.

Stage 4 – Tokenizing the summary

Tokenization breaks a text sequence into individual terms. Spark provides a simple `Tokenizer` class and a more flexible `RegexTokenizer` class, which we use here. The `pattern` parameter specifies a `"[!,\"  ]"` **regular expression** (**regex**) used to indicate how to split text into words. We split text on exclamation points, commas, quotes, and spaces (note the `\"` escaped quote in the regex), and we specify **input and output** (**I/O**) columns and force all words to lowercase, so that for example—`This` and `this` will be considered identical terms upon later processing. The code is illustrated in the following snippet:

```
from pyspark.ml.feature import RegexTokenizer
regexTokenizer = RegexTokenizer(inputCol = "Summary",
                                outputCol = "Tokenized",
                                pattern = "[!,\"]",
                                toLowercase = True)
```

Inspect the tokenized values from Summary, as follows:

```
tokenized = regexTokenizer.transform(created)
tokenized.select(["Tokenized"]).show(n = 10,
    truncate = False)
```

The output is shown in the following screenshot:

```
+----------------------------------------------------+
|Tokenized                                           |
+----------------------------------------------------+
|[good, quality, dog, food]                          |
|[not, as, advertised]                               |
|[delight, says, it, all]                            |
|[cough, medicine]                                   |
|[great, taffy]                                      |
|[nice, taffy]                                       |
|[great, just, as, good, as, the, expensive, brands]|
|[wonderful, tasty, taffy]                           |
|[yay, barley]                                       |
|[healthy, dog, food]                                |
+----------------------------------------------------+
only showing top 10 rows
```

Figure 6.9 – Results from the regexTokenizer stage 4 transformer

Phrases have now been broken up into lists of individual terms or tokens. Since our goal is to extract information from these tokens, we next filter out words that carry little information.

Stage 5 – Removing stop words

Some words occur so frequently in language that they have very little predictive value. These are termed *stop words* and we use Spark's StopWordsRemover transformer to delete them, as illustrated in the following screenshot:

```
removeStopWords = StopWordsRemover(
    inputCol = regexTokenizer.getOutputCol(),
    outputCol = "CleanedSummary",
    caseSensitive = False)
```

Let's compare the tokenized results before and after removing stop words, as follows:

```
stopWordsRemoved = removeStopWords.transform(tokenized)
stopWordsRemoved.select(["Tokenized",
                         "CleanedSummary"]).show(
    n = 10, truncate = False)
```

The results are displayed in the following screenshot:

```
+-------------------------------------------------------+-------------------------------------+
|Tokenized                                              |CleanedSummary                       |
+-------------------------------------------------------+-------------------------------------+
|[good, quality, dog, food]                             |[good, quality, dog, food]           |
|[not, as, advertised]                                  |[advertised]                         |
|[delight, says, it, all]                               |[delight, says]                      |
|[cough, medicine]                                      |[cough, medicine]                    |
|[great, taffy]                                         |[great, taffy]                       |
|[nice, taffy]                                          |[nice, taffy]                        |
|[great, just, as, good, as, the, expensive, brands]    |[great, good, expensive, brands]     |
|[wonderful, tasty, taffy]                              |[wonderful, tasty, taffy]            |
|[yay, barley]                                          |[yay, barley]                        |
|[healthy, dog, food]                                   |[healthy, dog, food]                 |
+-------------------------------------------------------+-------------------------------------+
only showing top 10 rows
```

Figure 6.10 – Results from the removeStopWords stage 5 transformer

Inspecting the results in *Figure 6.10* is illustrative. For the most part, removing stop words such as as, it, or the has little effect on meaning: the *Great! Just as good as the expensive brands!* statement being reduced to tokens [great, good, expensive, brands] seems reasonable. But what about *Not as advertised!* being reduced to [advertised]? The not in the statement would seem to carry important information that is lost by its removal. This is a valid concern that could be addressed by NLP concepts such as n-grams (bigrams, trigrams, and so on). For an example demonstrating Sparkling Water pipelines, we will acknowledge this as a potential issue but move on for simplicity.

NLP in predictive modeling represents information in text as numbers. A popular approach is **term frequency-inverse document frequency (TF-IDF)**. TF is simply the number of times a term appears in a document divided by the number of words in a document. In a corpus (collection of documents), IDF measures how rare a term is across its constituent documents. A term such as *linear* may have high frequency, but its information value decreases as the number of documents it appears in increases. On the other hand, a word such as *motorcycle* may have lower frequency but also be found in fewer documents in a corpus, making its information content higher. Multiplying TF by IDF gives a rescaled TF value that has proven quite useful. TF-IDF values are maximized when a term is found frequently but only in one document (*Which article reviewed motorcycles?*).

TF-IDF is widely used in information retrieval, text mining, recommender systems, and search engines, as well as in predictive modeling. The next two pipeline stages will compute the TF and IDF values, respectively.

Stage 6 – Hashing words for TF

Our preferred way to compute TF in Spark is CountVectorizer, which preserves the mapping from the index back to the word using an internal vocabulary. That is, countVectorizerModel.vocabulary[5] looks up the word stored in index 5.

A trick for building better TF-IDF models is to remove terms that are too infrequent by setting the minDF parameter as an integer or proportion, as follows:

- minDF = 100: Omit terms that appear in fewer than 100 documents

- minDF = 0.05: Omit terms that appear in fewer than 5% of documents

A maxDF parameter is also available for removing terms that occur too frequently across a corpus. For instance, setting maxDF = 0.95 in NLP for document retrieval might improve model performance.

We create a countVectorizer transformer, as follows:

```
from pyspark.ml.feature import CountVectorizer
countVectorizer = CountVectorizer(
    inputCol = removeStopWords.getOutputCol(),
    outputCol = "frequencies",
    minDF = 100 )
```

Note that our corpus is the output column of the removeStopWords transformer, with each row as a document. We output the frequencies and set minDF to 100. Because countVectorizer is a model, it is a good idea to manually train it before executing it in the pipeline. This is a good practice for any model that is a pipeline component as it allows us to determine its behavior and perhaps fine-tune it before pipeline execution commences. The code is illustrated in the following snippet:

```
countVecModel = countVectorizer.fit(stopWordsRemoved)
```

We can explore this model by inspecting its vocabulary size and individual terms, as well as any other appropriate due diligence. Here's the code we'd need to accomplish this:

```
print("Vocabulary size is " +
    str(len(countVecModel.vocabulary)))
print(countVecModel.vocabulary[:7])
```

The vocabulary results are shown here:

```
Vocabulary size is 1431
['great', 'good', 'best', 'love', 'coffee', 'tea', 'delicious']
```

Figure 6.11 – Vocabulary size and vocabulary of the countVecModel transformer

Our total vocabulary size shown in *Figure 6.11* is 1,431 terms. Inspect the data with the following code:

```
vectorized = countVecModel.transform(stopWordsRemoved)
vectorized.select(["CleanedSummary", "frequencies"]).show(
                    n = 10, truncate = False)
```

The vectorized result is shown in the following screenshot:

```
+---------------------------------+-----------------------------------------+
|CleanedSummary                   |frequencies                              |
+---------------------------------+-----------------------------------------+
|[good, quality, dog, food]       |(1431,[1,10,11,38],[1.0,1.0,1.0,1.0])    |
|[advertised]                     |(1431,[565],[1.0])                       |
|[delight, says]                  |(1431,[397,415],[1.0,1.0])               |
|[cough, medicine]                |(1431,[],[])                             |
|[great, taffy]                   |(1431,[0,1285],[1.0,1.0])                |
|[nice, taffy]                    |(1431,[31,1285],[1.0,1.0])               |
|[great, good, expensive, brands] |(1431,[0,1,112,807],[1.0,1.0,1.0,1.0])|
|[wonderful, tasty, taffy]        |(1431,[15,30,1285],[1.0,1.0,1.0])        |
|[yay, barley]                    |(1431,[1293],[1.0])                      |
|[healthy, dog, food]             |(1431,[10,11,24],[1.0,1.0,1.0])          |
+---------------------------------+-----------------------------------------+
only showing top 10 rows
```

Figure 6.12 – Intermediate results from the countVecModel transformer

Figure 6.12 shows the cleaned summary tokens with TF vectors side by side for each row. To describe the output for the first row, the 1431 value is the vocabulary size. The next sequence of values— [1,10,11,38] —refers to the indices of the [good, quality, dog, food] terms in the vocabulary vector. The last series of values— [1.0,1.0,1.0,1.0] —are the TFs for their respective terms. Thus, dog is referenced by index 11 and occurs once in the CleanedSummary column.

Stage 7 – Creating an IDF model

We use Spark's IDF estimator to scale frequencies from countVectorizer, yielding TF-IDF values. We execute the following code to accomplish this:

```
from pyspark.ml.feature import IDF
idf = IDF(inputCol = countVectorizer.getOutputCol(),
```

```
            outputCol = "TFIDF",
            minDocFreq = 1)
```

Manually train the IDF model to see the results before we execute the pipeline, like so:

```
idfModel = idf.fit(vectorized)
```

Inspect the data again, noting especially the scaled TF-IDF frequencies, as follows:

```
afterIdf = idfModel.transform(vectorized)
afterIdf.select(["Sentiment", "CleanedSummary",
    "TFIDF"]).show(n = 5, truncate = False, vertical = True)
```

The first five rows of the resulting TF-IDF model are shown in the following screenshot:

```
-RECORD 0-------------------------------------------------------------------------------------------
 Sentiment      | Positive
 CleanedSummary | [good, quality, dog, food]
 TFIDF          | (1431,[1,10,11,38],[2.5111924617383754,3.7820835954087735,3.7893082516106595,4.670183735977007])
-RECORD 1-------------------------------------------------------------------------------------------
 Sentiment      | Negative
 CleanedSummary | [advertised]
 TFIDF          | (1431,[565],[7.222887042775855])
-RECORD 2-------------------------------------------------------------------------------------------
 Sentiment      | Positive
 CleanedSummary | [delight, says]
 TFIDF          | (1431,[397,415],[6.871606368620513,6.880554974196527])
-RECORD 3-------------------------------------------------------------------------------------------
 Sentiment      | Negative
 CleanedSummary | [cough, medicine]
 TFIDF          | (1431,[],[])
-RECORD 4-------------------------------------------------------------------------------------------
 Sentiment      | Positive
 CleanedSummary | [great, taffy]
 TFIDF          | (1431,[0,1285],[2.073077026932758,8.233697127999516])
only showing top 5 rows
```

Figure 6.13 – TF-IDF frequencies from the Spark transformer

Stage 8 – Selecting modeling dataset columns

In addition to the Sentiment response variable and all the features engineered from the Time variable, the output of idf includes the original Summary column as well as Tokenized, CleanedSummary, frequencies, and TFIDF. Of these, we wish to keep only TFIDF. The following code selects the desired columns:

```
finalSelect = SQLTransformer(
    statement="""
    SELECT Sentiment, Day, Month, Year, WeekNum, Weekday,
           HourOfDay, Weekend, Season, TFIDF
    FROM __THIS__ """)
```

Now that we are finished building our model-ready data, the next step is to build a predictive model using one of H2O's supervised learning algorithms.

Stage 9 – Creating an XGBoost model using H2O

Up to this point, all our data wrangling and feature engineering efforts have used Spark methods exclusively. Now, we turn to H2O to train an XGBoost model on the Sentiment column. For simplicity, we train using default settings. The code is illustrated in the following snippet:

```
import h2o
from pysparkling.ml import ColumnPruner, H2OXGBoost
xgboost = H2OXGBoost(splitRatio = 0.8, labelCol = "Sentiment")
```

> **Note – Training Models in Sparkling Water**
>
> In *Chapter 5*, *Advanced Model Building – Part I*, we demonstrated in detail a process for building and tuning a high-quality XGBoost model. We stop at a simple baseline model here to emphasize the utility of the overall pipeline. In a real application, much more effort should be spent on the modeling component of this pipeline.

Creating a Sparkling Water pipeline

Now that we have all the transformers defined, we are ready to create a pipeline. Doing so is simple—we just name each transformer in order in the stages list parameter of Pipeline, as follows:

```
from pyspark.ml import Pipeline
pipeline = Pipeline(stages = [
    colSelect, expandTime, createResponse, regexTokenizer,
    removeStopWords, countVectorizer, idf, finalSelect,
    xgboost])
```

Training the pipeline model is made simple by using the fit method. We pass as a parameter the Spark DataFrame containing the raw data, as follows:

```
model = pipeline.fit(reviews_spark)
```

During the pipeline.fit process, the data munging and feature engineering stages are all applied to the raw data in the order defined before the XGBoost model is fit. These pipeline stages operate identically after deployment in production with the XGBoost stage, producing predictions.

Looking ahead – a production preview

Putting the Sparkling Water pipeline into production is simply a matter of *saving* the `pipeline` model, *loading* it onto the production system, then calling the following:

```
predictions = model.transform(input_data)
```

In *Chapter 10, H2O Model Deployment Patterns*, we show how to deploy this pipeline as a Spark streaming application, with the pipeline receiving raw streaming data and outputting predictions in real time.

UL methods in H2O

H2O includes several unsupervised learning algorithms including **Generalized Low Rank Models (GLRM)**, **Principal Component Analysis (PCA)**, and an aggregator for dimensionality reduction. Clustering use cases can utilize k-means clustering, H2O aggregator, GLRM, or PCA. Unsupervised learning also underlies a set of useful feature transformers used in predictive modeling applications—for example, the distance of an observation to a specific data cluster identified by an unsupervised method. In addition, H2O provides an isolation forest algorithm for anomaly detection.

What is anomaly detection?

Most **machine learning (ML)** algorithms attempt, in some manner, to find patterns in data. These patterns are leveraged to make predictions in supervised learning models. Many unsupervised learning algorithms try to uncover patterns through clustering similar data or estimating boundaries between data segments. Unsupervised anomaly detection algorithms take the opposite approach: data points that do not follow known patterns are what we want to discover.

In this context, the term *anomaly* is value-free. It may refer to an unusual observation because it is the first of its kind; more data could yield additional similar observations. Anomalies might be indicative of unexpected events and serve as a diagnostic. For instance, a failed sensor in a manufacturing data-collection application could yield atypical measurements. Anomalies may also indicate malicious actors or actions: security breaches and fraud are two classic examples resulting in anomalous data points.

Anomaly detection approaches may include supervised, semi-supervised, or unsupervised methods. Supervised models are the gold standard in fraud detection. However, obtaining labels for each observation can be costly and is often not feasible. An unsupervised approach is required when labels are absent. Semi-supervised approaches refer to situations where only some of the data records are labeled, usually a small minority of records.

Isolation forest is a unsupervised learning algorithm for anomaly detection—we'll introduce this next.

Isolation forests in H2O

The isolation forest algorithm is based on decision trees and a clever observation: outliers tend to be split out very early in the building of a decision tree. But decision trees are a supervised method, so how is this unsupervised? The trick is to create a target column of random values and train a decision tree on it. We repeat this many times and record the average depth at which observations are split into their own leaf. The earlier an observation is isolated, the more likely it is to be anomalous. Depending on the use case, these anomalous points may be filtered out or escalated for further investigation.

You can see a representation of an isolation forest in the following screenshot:

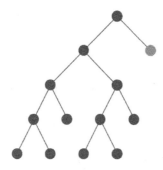

Figure 6.14 – An isolation forest

We show how to build an isolation forest in H2O using the Kaggle credit card transaction data (`https://www.kaggle.com/mlg-ulb/creditcardfraud`). There are 492 fraudulent and 284,807 non-fraudulent transactions in this dataset, which makes the target class highly imbalanced. Because we are demonstrating an unsupervised anomaly detection approach, we will drop the labeled target during the model build.

The H2O code for loading the data is shown here:

```
df = h2o.import_file("creditcardfraud.csv")
```

We fit our isolation forest using the `H2OIsolationForestEstimator` method. We set the number of trees to `100` and omit the last column, which contains the target class label, as shown in the following code snippet:

```
iso = h2o.estimators.H2OIsolationForestEstimator(
    ntrees = 100, seed = 12345)
iso.train(x = df.col_names[0:31], training_frame = df)
```

Once the model is trained, prediction is straightforward, as we can see here:

```
predictions = iso.predict(df)
predictions
```

The output is shown in the following screenshot:

predict	mean_length
0.0810056	6.71
0.0418994	6.85
0.162011	6.42
0.120112	6.57
0.0614525	6.78
0.047486	6.83
0.0418994	6.85
0.240223	6.14
0.0726257	6.74
0.0502793	6.82

Figure 6.15 – Isolation forest predictions

The predictions in *Figure 6.15* consist of two columns: the normalized anomaly score and the average number of splits across all trees to isolate the observation. Note that the anomaly score is perfectly correlated with mean length, increasing as mean length decreases.

How do we go from either an anomaly score or mean length to an actual prediction? One of the best ways is through a quantile-based threshold. If we have an idea about the prevalence of fraud, we can find the corresponding quantile value of the score and use it as a threshold for our predictions. Suppose we know that 5% of our transactions are fraudulent. Then, we estimate the correct quantile using the following H2O code:

```
quantile = 0.95
quantile_frame = predictions.quantile([quantile])
quantile_frame
```

The resulting quantile output is shown in the following screenshot:

Probs	predictQuantiles	mean_lengthQuantiles
0.95	0.198324	6.99

Figure 6.16 – Choosing a quantile-based threshold

We now can use the threshold to predict the anomalous class using the following code:

```
threshold = quantile_frame[0, "predictQuantiles"]
predictions["predicted_class"] = \
    predictions["predict"] > threshold
predictions["class"] = df["class"]
predictions
```

The first 10 observations of the predictions frame are shown in the following screenshot:

predict	mean_length	predicted_class	class
0.0810056	6.71	0	0
0.0418994	6.85	0	0
0.162011	6.42	0	0
0.120112	6.57	0	0
0.0614525	6.78	0	0
0.047486	6.83	0	0
0.0418994	6.85	0	0
0.240223	6.14	1	0
0.0726257	6.74	0	0
0.0502793	6.82	0	0

Figure 6.17 – Identifying anomalous values

The predict column in *Figure 6.17* has only one observation greater than 0.198324, the threshold for the 95th percentile shown in *Figure 6.16*. The predicted_class column indicates this with a value of 1. Also, note that the mean_length value for this observation of 6.14 is less than the mean length values for the other nine observations.

The class column contains the transaction fraud indicator that we omitted in building the unsupervised isolation forest model. For the anomalous observation, a class value of 0 indicates that the transaction was not fraudulent. When we have access to an actual target value as in this example, we could use the predicted_class and class columns to study the effectiveness of the anomaly detection algorithm in detecting fraud. We should note that the definition of fraud and anomalous in this context are not equivalent. In other words, not all frauds are anomalous and not all anomalies will indicate fraud. These two models have separate, albeit complementary, purposes.

We now turn our attention to updating models.

Best practices for updating H2O models

As the famous British statistician George Box stated, *All models are wrong, but some are useful.* Good modelers are aware of the purpose as well as the limitations of their models. This should be especially true for those who build enterprise models that go into production.

One such limitation is that predictive models as a rule degrade over time. This is largely because, in the real world, things change. Perhaps what we are modeling—customer behavior, for example—itself changes, and the data we collect reflects that change. Even if customer behavior is static but our mix of business changes (think more teenagers and fewer retirees), our model's predictions will likely degrade. In both cases but for different reasons, the population that was sampled to create our predictive model is not the same now as it was before.

Detecting model degradation and searching for its root cause is the subject of diagnostics and model monitoring, which we do not address here. Rather, once a model is no longer satisfactory, what should a data scientist do? We address retraining and checkpointing models in the following sections.

Retraining models

Developing a parametric model entails:

1. Finding the correct structural form for the process being modeled and then

2. Using data to estimate the parameters of that structure. Over time, if the structure of the model remains the same but the data changes, then we could *refit* (or *retrain* or *re-estimate* or *update*) the model's parameter estimates. This is a simple and relatively straightforward procedure.

However, if the underlying process changes in a way that means the structural form of the model is no longer valid, then modeling consists of both discovering the correct form of the model and estimating parameters. This is almost, but not quite, the same thing as starting from scratch. *Rebuilding* or *updating the model* (as opposed to *updating the parameter estimates*) are better terms for this larger activity.

In the case of ML or other nonparametric models, the structural form of the model is determined by the data along with any parameter estimates. This is one of the selling points of nonparametric models: they are incredibly data-driven and nearly assumption-free. The differences between refitting or retraining and rebuilding have little meaning in this context; these terms become, in effect, synonyms.

Checkpointing models

The **Checkpoint** option in H2O lets you save the state of a model build, allowing a new model to be built as a *continuation* of a previously generated model rather than building one from scratch. This can be used to update a model in production with additional, more current data.

The checkpoint option is available for **Distributed Random Forest (DRF)**, **Gradient Boosting Machine (GBM)**, XGBoost, and **deep learning (DL)** algorithms. For tree-based algorithms, the number of trees specified must be greater than the number of trees in the referenced model. That is, if the original model included 20 trees and you specify 30 trees, then 10 new trees will be built. The same concept is true for DL using epochs rather than trees.

Checkpointing is feasible for these algorithms *only* when the following are the same as the checkpointed model:

- The training data model type, response type, columns, categorical factor levels, and the total number of predictors
- The identical validation dataset if one was used in the checkpointed model (cross-validation is not currently supported for checkpointing)

Additional parameters that you can specify with checkpointing vary based on the algorithm that was used for model training.

> **Checkpointing Caveats**
>
> Although it is technically feasible, we do not recommend checkpointing on new data for GBM or XGBoost algorithms. Recall that boosting works by fitting sequential models to the residuals of previous models. The early splits are thus the most important. By the time new data has been introduced, the structure of the model has been largely determined in its absence.
>
> Checkpointing **random forest** models does not suffer from these concerns due to differences between boosting and bagging.

Ensuring H2O model reproducibility

In a laboratory or experimental setting, repeating a process under the same protocols and conditions should lead to similar results. Natural variability may of course occur, but this can be measured and attributed to appropriate factors. This is termed *repeatability*. The enterprise data scientist should ensure that their model builds are well coded and sufficiently documented to make the process repeatable.

Reproducibility in the context of model building is a much stronger condition: the results when a process is repeated must be identical. From a regulatory or compliance perspective, reproducibility may be required.

At a high level, reproducibility requires the same hardware, software, data, and settings. Let's review this specifically for H2O setups. We begin with two cases depending on the H2O cluster type.

Case 1 – Reproducibility in single-node clusters

A single-node cluster is the simplest H2O hardware configuration. Reproducibility can be attained if the following conditions are met:

- **Software requirements**: The same version of H2O-3 or Sparkling Water is used.

- **Data requirements**: The same training data is used (note that H2O requires files to be imported individually rather than as an entire directory to guarantee reproducibility).

- **Settings requirements**:

  - The same parameters are used to train the model.

  - If sampling is done in the algorithm, the same seed is required. This includes any of the following: `sample_rate`, `sample_rate_per_class`, `col_sample_rate`, `col_sample_rate_per_level`, `col_sample_rate_per_tree`.

  - If early stopping is enabled, reproducibility is only guaranteed when the `score_tree_interval` parameter is explicitly set and the same validation dataset is used.

Case 2 – Reproducibility in multi-node clusters

Adding nodes to a cluster creates additional hardware conditions that must be met in order to achieve reproducibility. The software, data, and settings requirements are the same as in single-node clusters detailed previously in *Case 1*. These requirements are outlined here:

- The hardware cluster must be configured identically. Specifically, clusters must have the same number of nodes with the same number of CPU cores per node or—alternatively—the same restriction on the number of threads.

- The cluster's leader node must initiate model training. In Hadoop, the leader node is automatically returned to the user. In standalone deployments, the leader node must be manually identified. See the H2O documentation for more details.

For reproducibility, you must ensure that the cluster configuration is identical. The parallelization level (number of nodes and CPU cores/threads) controls how a dataset is partitioned in memory. H2O runs its tasks in a predictable order on these partitions. If the number of partitions is different, the results will not be reproducible.

In cases where the cluster configuration is not identical, it may be possible to constrain the resources of computations being reproduced. This process involves replicating data partitions in the original environment. We refer you to the H2O documentation for more information.

Reproducibility for specific algorithms

The complexity of DL, GBM, and **automated ML (AutoML)** algorithms introduces additional constraints that must be met in order to ensure reproducibility. We will review these requirements in this section.

DL

H2O DL models are not reproducible by default for performance reasons. There is a `reproducible` option that can be enabled, but we recommend doing this only for small data. The model takes significantly more time to generate because only one thread is used for computation.

GBM

GBM is deterministic up to floating-point rounding errors when reproducibility criteria are met for single- or multi-node clusters.

AutoML

To ensure reproducibility in AutoML, the following criteria must be met:

- DL must be excluded.
- The `max_models` constraint rather than `max_runtime_secs` must be used.

As a rule, time-based constraints are resource-limited. This means that AutoML may be able to train more models on one run than another if available compute resources differ between runs. Specifying the number of models to build will ensure reproducibility.

Best practices for reproducibility

To ensure reproducibility, think of the four requirement categories emphasized earlier: hardware, software, data, and settings. These categories are explained in more detail here:

- **Hardware**: You should always document the hardware resources the H2O cluster is running on—this includes the number of nodes, CPU cores, and threads. (This information can be found in the log files.)

- **Software**: You should document the version of H2O-3 or Sparkling Water used. (This information can be found in the log files.)

- **Data**: Obviously, you must use the same input data. You should save all scripts that were used to process the data prior to model training. All data column modifications should be documented (for example, if you converted a numeric column to a categorical one).

- **Settings**: Save the H2O logs and the H2O binary model. The logs contain a wealth of information. More importantly, the binary model contains the H2O version (software) and the parameters used to train the model (settings).

Summary

In this chapter, we have rounded out our advanced modeling topics by showing how to build H2O models in Spark pipelines with a hands-on sentiment analysis modeling example. We summarized the unsupervised learning methods available in H2O and showed how to build an anomaly detection model using the isolation forest algorithm for a credit card fraud transaction use case. We also reviewed how to update models, including refitting versus checkpointing, and showed requirements to ensure model reproducibility.

In *Chapter 7, Understanding ML Models,* we discuss approaches for understanding and reviewing our ML models.

7
Understanding ML Models

Now that we have built a few models using H2O software, the next step before production is to understand how the model is making decisions. This has been termed variously as **machine learning interpretability (MLI)**, **explainable artificial intelligence (XAI)**, model explainability, and so on. The gist of all these terms is that building a model that predicts well is not enough. There is an inherent risk in deploying any model before fully trusting it. In this chapter, we outline a set of capabilities within H2O for explaining ML models.

By the end of this chapter, you will be able to do the following:

- Select an appropriate model metric for evaluating your models.

- Explain what Shapley values are and how they can be used.

- Describe the differences between global and local explainability.

- Use multiple diagnostics to build understanding and trust in a model.

- Use global and local explanations along with model performance metrics to choose the best among a set of candidate models.

- Evaluate tradeoffs between model predictive performance, speed of scoring, and assumptions met in a single candidate model.

In this chapter, we're going to cover the following main topics:

- Selecting model performance metrics
- Explaining models built in H2O (both globally and locally)
- Automated model documentation through H2O AutoDoc

Selecting model performance metrics

The most relevant question about any model is, *How well does it predict?* Regardless of any other positive properties that a model may possess, models that don't predict well are just not very useful. How to best measure predictive performance depends both on the specific problem being solved and the choices available to the data scientist. H2O provides multiple options for measuring model performance.

For measuring predictive model performance in regression problems, H2O provides R^2, **mean squared error (MSE)**, **root mean squared error (RMSE)**, **root mean squared logarithmic error (RMSLE)**, and **mean absolute error (MAE)** as metrics. MSE and RMSE are good default options, with RMSE being our preference because the metric is expressed in the same units as the predictions (rather than squared units, as in the case of MSE). All metrics based on squared error are sensitive to outliers in general. If robustness to outliers is a requirement, then MAE is a better choice. Finally, RMSLE is useful in the special case where under-prediction is worse than over-prediction.

For classification models, H2O adds the Gini coefficient, absolute **Matthews correlation coefficient (MCC)**, F1, F0.5, F2, Accuracy, Logloss, **area under the ROC curve (AUC)**, **area under the precision-recall curve (AUCPR)**, and **Kolmogorov-Smirnov (KS)** metrics. In our experience, AUC is the most commonly used metric in business. Because communication with business partners and executives is so vital to data scientists, we recommend using well-known metrics when their use is appropriate for the job. In the case of AUC, it does a good job with binary classification models when data is relatively balanced. AUCPR is a better choice for imbalanced data.

TheLogloss metric, based on information theory, has some mathematical advantages. In particular, if you are interested in the predicted probabilities of class membership themselves and not just the predicted classifications, Logloss is a better choice of metric. Further documentation on these scoring options can be found at `https://docs.h2o. ai/h2o/latest-stable/h2o-docs/performance-and-prediction.html`.

Leaderboards created in AutoML for a classification problem include AUC, Logloss, AUCPR, mean per-class error, RMSE, and MSE as performance metrics. The leaderboard for the `check` AutoML object created in *Chapter 5, Advanced Model Building – Part 1*, is shown in *Figure 7.1* as an example:

model_id	auc	logloss	aucpr	mean_per_class_error	rmse	mse	training_time_ms	predict_time_per_row_ms	algo
StackedEnsemble_AllModels_AutoML_20211011_125447	0.710969	0.402348	0.325487	0.34709	0.351922	0.123849	1044	0.024207	StackedEnsemble
StackedEnsemble_BestOfFamily_AutoML_20211011_125447	0.708667	0.403303	0.323737	0.348399	0.352255	0.124083	786	0.010713	StackedEnsemble
GBM_1_AutoML_20211011_125447	0.704856	0.405453	0.316492	0.351304	0.353352	0.124858	550	0.003996	GBM
GBM_2_AutoML_20211011_125447	0.701839	0.406387	0.314821	0.358446	0.353761	0.125147	604	0.004059	GBM
GBM_3_AutoML_20211011_125447	0.701713	0.40697	0.313242	0.357351	0.354126	0.125405	672	0.004579	GBM
GBM_5_AutoML_20211011_125447	0.70133	0.406057	0.317528	0.359195	0.353353	0.124858	935	0.004571	GBM
XGBoost_3_AutoML_20211011_125447	0.697141	0.408385	0.311164	0.358829	0.354663	0.125786	785	0.001189	XGBoost
GBM_4_AutoML_20211011_125447	0.693177	0.411546	0.305102	0.35777	0.356183	0.126866	792	0.00414	GBM
GLM_1_AutoML_20211011_125447	0.690928	0.409483	0.306502	0.365288	0.354481	0.125657	1019	0.001031	GLM
DRF_1_AutoML_20211011_125447	0.687825	0.415476	0.300201	0.364479	0.356173	0.126859	2298	0.006808	DRF
XGBoost_1_AutoML_20211011_125447	0.674599	0.428943	0.285501	0.374881	0.362641	0.131508	1902	0.001464	XGBoost
XGBoost_2_AutoML_20211011_125447	0.670094	0.434783	0.283699	0.378092	0.364279	0.132899	2301	0.001349	XGBoost

Figure 7.1 – An AutoML leaderboard for the check object

In addition to predictive performance, additional metrics of model performance may be important in an enterprise setting. Two of these included by default in AutoML leaderboards are the amount of time required to fit a model (`training_time_ms`) and the amount of time required to predict a single row of the data (`predict_time_per_row_ms`).

In *Figure 7.1*, the best model according to both AUC and Logloss is a stacked ensemble of all models (the model in the top row). This model is also the slowest to score by an order of magnitude over any of the individual models. For streaming or real-time applications in particular, a model that cannot score quickly enough may automatically be disqualified as a candidate regardless of its predictive performance.

We next address model explainability for understanding and evaluating our ML models.

Explaining models built in H2O

Model performance metrics measured on our test data can tell us how well a model predicts and how fast it predicts. As mentioned in the chapter introduction, knowing that a model predicts well is not a sufficient reason to put it into production. Performance metrics alone cannot provide any insight into *why* the model is predicting as it is. If we don't understand why the model is predicting well, we have little hope of being able to anticipate conditions that would make the model not work well. The ability to explain a model's reasoning is a critical step prior to promoting it into production. This process can be described as gaining trust in the model.

Explainability is typically divided into global and local components. Global explainability describes how the model works for an entire population. Gaining trust in a model is primarily a function of determining how it works globally. Local explanations operate instead on individual rows. They address questions such as how an individual prediction came about. The `h2o.explain` and `h2o.explain_row` methods bundle a set of explainability functions and visualizations for global and local explanations, respectively.

We start this section with a simple introduction to Shapley values, one of the bedrock methods in model explainability, which can be confusing when first encountered. We cover global explanations for single models using `h2o.explain` and local explainability with `h2o.explain_row`. We then address global explanations for AutoML using `h2o.explain`, which we use to demonstrate the role of explainability in model selection. We illustrate the output of these methods using two models developed in *Chapter 5, Advanced Model Building – Part 1*. The first, `gbm`, is an individual baseline model built using default values with the H2O **Gradient Boosting Machine (GBM)** estimator. The second is an AutoML object, `check`. These models are chosen as examples only, acknowledging that the original baseline model was improved upon by multiple feature engineering and model optimization steps.

A simple introduction to Shapley values

Shapley values have become an important part of ML explainability as a means for attributing the contribution of each feature to either overall or individual predictions. Shapley values are mathematically elegant and well-suited for the task of attribution. In this section, we provide a description of Shapley values: their origin, calculation, and how to use them for interpretation.

Lloyd Shapley (1923-2016), a 2012 Nobel Prize winner in Economics, derived Shapley values in 1953 as the solution to a specific problem in game theory. Suppose a group of players working together receives a prize. How should that award be equitably divided amongst the players?

Shapley started with mathematical axioms defining fairness: **symmetry** (players who contribute the same amount get the same payout), **dummy** (players who contribute nothing receive nothing), and **additivity** (if the game can be separated into additive parts, then you can decompose the payouts). The Shapley value is the unique mathematical solution that satisfies these axioms. In short, the Shapley value approach pays players in proportion to their marginal contributions.

We next demonstrate the calculation of Shapley values for a couple of simple scenarios.

Shapley calculations illustrated – Two players

To illustrate the calculation of a Shapley value, consider the following simple example. Two musicians, John and Paul, performing on their own can earn £4 and £3, respectively. John and Paul playing together earn £10. How should they divide the £10?

To calculate their marginal contributions, consider the number of ways these players can be sequenced. For two players, there are only two unique orderings: John is playing and is then joined by Paul, or Paul is playing and is then joined by John. This is illustrated in *Figure 7.2*:

1.	John → John and Paul
2.	Paul → John and Paul

Figure 7.2 – Unique player sequences for John and Paul

The formulation as unique player sequences allows us to calculate Shapley values for each player. We illustrate the calculation of the Shapley value for John in *Figure 7.3*:

1.	John → John and Paul	v(J)			=	4
2.	Paul → John and Paul	v(JP) – v(P)	=	10 - 3	=	7
						11

Figure 7.3 – Sequence values for John

John is the first player present in sequence 1, thus the Shapley contribution is just the marginal value $v(J) = 4$. In the second sequence, John joins after Paul. The marginal value for John is the joint value of John and Paul, $v(JP)$, minus the marginal value of Paul, $v(P)$. In other words, $10 – 3 = 7$. The Shapley value for John is the average of the values for each sequence: $S(J) = 11/2 = 5.5$. Therefore, John should receive £5.50 of the £10 payment.

The Shapley value for Paul is calculated in a similar fashion (obviously, it could also be calculated by subtraction). The sequence calculations are shown in *Figure 7.4*:

1.	John → John and Paul	v(JP) – v(J)	=	10 - 4	=	6
2.	Paul → John and Paul	v(P)			=	3
						9

Figure 7.4 – Sequence values for Paul

In the first sequence in *Figure 7.4*, Paul joins after John, so the sequence value is the joint, $v(JP) = 10$, minus the marginal for John, $v(J) = 4$. The second sequence is just the marginal value of Paul: $v(P)=3$. The Shapley value for Paul is $S(P) = 9/2 = 4.5$.

These calculations are easy and make sense with two players. Let's see what happens when we add a third player.

Shapley calculations illustrated – Three players

Suppose a third musician, George, joins John and Paul. George earns £2 on his own, £7 performing with John, £9 performing with Paul, and £20 when all three play together. For clarity, we summarize the earnings in *Figure 7.5*:

John	£4	John and Paul	£10	John and Paul and George	£20
Paul	£3	John and George	£7		
George	£2	Paul and George	£9		

Figure 7.5 – Earnings for John, Paul, and George

Because there are three players, there are 3! = 6 unique sequences in which John, Paul, and George can arrive. The calculations for the Shapley value for John in this three-player scenario are summarized in *Figure 7.6*:

1.	J	→	JP	→	JPG	v(J)			=	4
2.	J	→	JG	→	JPG	v(J)			=	4
3.	P	→	JP	→	JPG	v(JP) – v(P)	=	10 – 3	=	7
4.	P	→	PG	→	JPG	v(JPG) – v(PG)	=	20 – 9	=	11
5.	G	→	JG	→	JPG	v(JG) – v(G)	=	7 – 2	=	5
6.	G	→	PG	→	JPG	v(JPG) – v(PG)	=	20 – 9	=	11
									42

Figure 7.6 – Arrival sequences and values for calculating the Shapley value for John

In *Figure 7.6*, sequences 1 and 2 are straightforward: John is the first player, so the *v(J)* value is all that is needed. In sequences 3 and 5, John is the second player. The sequence values are calculated by taking the joint value of John and the first player and then subtracting the marginal value of that player. Sequences 4 and 6 are identical: John is the last player. His marginal contribution is calculated by taking the three-way interaction, *v(JPG)*, and subtracting the joint value of Paul and George, *v(PG)*. The Shapley value is *S(J) = 42/6 = 7*.

We could continue and find the Shapley values for Paul and George in the same manner.

Calculating Shapley values for N players

As you can see, Shapley value calculations can quickly become overwhelming as the number of players, N, increases. The Shapley sequence calculations depend on knowing the values for the main effects and all the interactions from two-way to N-way, as in *Figure 7.5*. In addition, there are $N!$ sequences to be solved. The computational task increases dramatically as the number of players increases.

In the context of a predictive model, each feature is a player and the prediction is the shared prize. We can use Shapley values to attribute the impact of each feature on the final prediction. With some models having dozens or hundreds or possibly, even more, features, computing Shapley values in the real world is non-trivial. Fortunately, a combination of modern computing and mathematical shortcuts for computing Shapley values for certain families of models makes Shapley calculations tenable.

Whether in simple examples as we have shown or in large complex ML models, the interpretation of Shapley values is the same.

We next turn our attention to global explanations for single models.

Global explanations for single models

We illustrate single model explanations using the baseline GBM model built in *Chapter 5, Advanced Model Building – Part 1*. We labeled this model gbm and documented its performance in *Figure 5.5* through to *Figure 5.10*.

The basic command for global explanations is as follows:

```
model_object.explain(test)
```

Here, `test` is the holdout test dataset used in model evaluation. Additional optional parameters include the following:

- `top_n_features`: An integer indicating how many columns to use in column-based methods such as **SHapley Additive exPlanations** (**SHAP**)) and **Partial Dependence Plots** (**PDP**). Columns are ordered according to variable importance. The default value is 5.

- `columns`: A vector of column names to use in column-based methods as an alternative to `top_n_features`.

- `include_explanations` or `exclude_explanations`: Respectively, include or exclude methods such as `confusion_matrix`, `varimp`, `shap_summary`, or `pdp`.

For a single classification model such as gbm, this command will display the confusion matrix, variable importance plot, SHAP summary plot, and partial dependence plots for the top five variables in order of importance.

We demonstrate this using our gbm model with the gbm.explain(test) command and discuss each display in turn.

The confusion matrix

The first output result is the confusion matrix, shown in *Figure 7.7*:

```
gbm.explain(test)
```

Confusion Matrix

Confusion matrix shows a predicted class vs an actual class.

gbm_baseline

Confusion Matrix (Act/Pred) for max f1 @ threshold = 0.24611616871629685:

		0	1	Error	Rate
0	0	17616.0	2087.0	0.1059	(2087.0/19703.0)
1	1	1716.0	2057.0	0.4548	(1716.0/3773.0)
2	Total	19332.0	4144.0	0.162	(3803.0/23476.0)

Figure 7.7 – Confusion matrix for the GBM baseline model

A nice feature of the `explain` method is that simple summary descriptions are provided for each display: **Confusion matrix shows a predicted class vs an actual class**. In *Figure 7.7*, the confusion matrix for gbm shows true negatives (17,616), false positives (2,087), false negatives (1,716), and true positives (2,057), along with a false positive rate (10.59%) and a false negative rate (45.48%).

The variable importance plot

The second visualization from the `explain` method is the variable importance plot, shown in *Figure 7.8*:

Variable Importance

The variable importance plot shows the relative importance of the most important variables in the model.

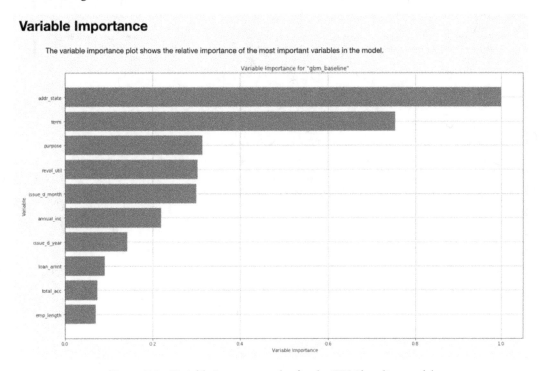

Figure 7.8 – Variable importance plot for the GBM baseline model

Note that the variable importance plot in *Figure 7.8* is identical to the plot displayed in *Figure 5.10* that we created manually using the `varimp_plot` command. Its inclusion here is one of the benefits of using the `explain` method.

The SHAP summary plot

The third visualization output by `explain` is a SHAP summary plot. **SHAP**, based on Shapley values, provides an informative view into black-box models. The SHAP summary plot for the GBM baseline model is shown in *Figure 7.9*:

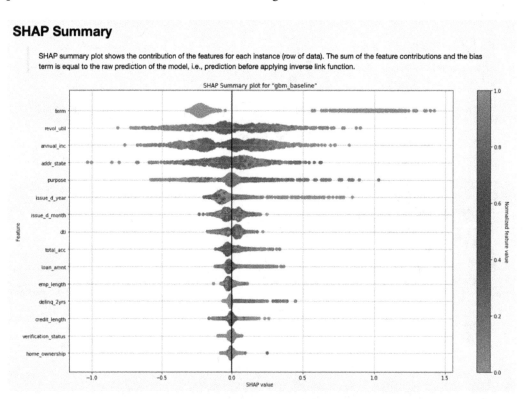

Figure 7.9 – SHAP summary plot for the GBM baseline model

Let's explain in a little more detail the SHAP summary plot in *Figure 7.9*. There is a lot going on in this informative plot:

- On the left-hand side, we have features (data columns) listed in order of decreasing feature importance based on Shapley values. (Note that Shapley feature importance rankings are not necessarily identical to the feature importance in *Figure 7.8*.)

- On the right-hand side, we have a normalized feature value scale going from **0.0** to **1.0** (blue to red as output by H2O). In other words, for each feature, we code the original data values by color: low original values as blue transitioning through purple for middling values, and ending with high original values as red (they show as varying shades of gray in this figure).

- The horizontal location of each observation is determined by its SHAP value. SHAP values measure the contribution of each feature to the prediction. Lower SHAP values are associated with lower predictions and higher values are associated with higher predictions.

With that initial understanding, we can make the following observations:

- Features that have red values to the right and blue values to the left are positively correlated with the response. Since we are modeling the probability of a bad loan, features such as longer term (`term`) or higher revolving utilization (`revol_util`) are positively correlated with loan default. (Revolving credit utilization is essentially how large a customer's credit card balances are month-to-month.)

- Features with red values to the left and blue values to the right are negatively correlated with the response. So, for example, higher annual income (`annual_inc`) is negatively correlated with a loan going into default.

These model observations from the SHAP summary plot make intuitive sense. You might expect someone who carries a larger credit card balance or makes a lower annual income to have an increased probability of defaulting on a loan.

Note that we can get the same plot using the `gbm.shap_summary_plot(test)` command.

Partial dependence plots

The fourth visualization output by `explain` is a set of partial dependence plots. The specific plots shown depend on the `top_n_features` or `columns` optional parameters. By default, the top five features are shown in order of decreasing variable importance. *Figure 7.10* displays the partial dependence plots for the address state:

Partial Dependence Plots

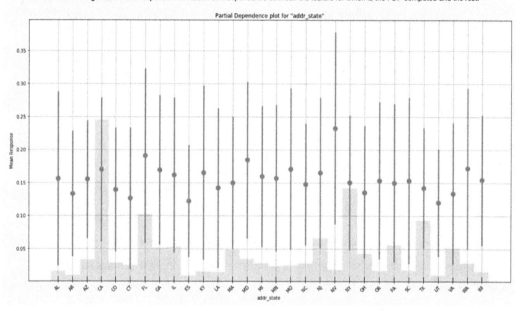

Figure 7.10 – Partial dependence plot for address state

Figure 7.11 displays the partial dependence plot for the revolving utilization variable:

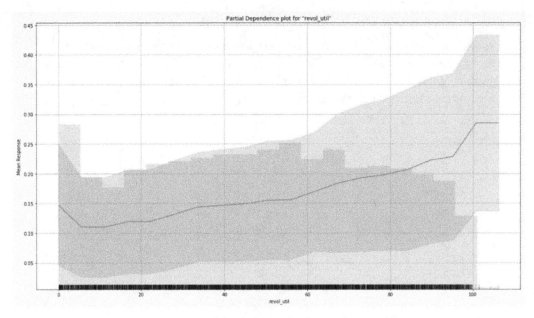

Figure 7.11 – Partial dependence plot for revolving utilization

The partial dependence plots output by explain include a representation of sample size (the shaded area starting from the bottom of the graph) overlayed by the mean response and its variability (the line surrounded by a shaded region). In the case of categorical variables, the mean response is a dot with bars indicating variability, as in *Figure 7.10*. In the case of numeric variables, the mean response is a dark line with lighter shading indicating variability, as in *Figure 7.11*.

Note that we can create a partial dependence plot for any individual column using the following:

```
gbm.pd_plot(test, column='revol_util')
```

The global individual conditional expectation (ICE) plot

ICE plots will be introduced later in the *Local explanations for single models* section. However, for completeness, we include a global version of the ICE plot here. Note that this plot is not output by `explain`. The `gbm.ice_plot(test, column='revol_util')` command returns a global ICE plot as shown in *Figure 7.12*:

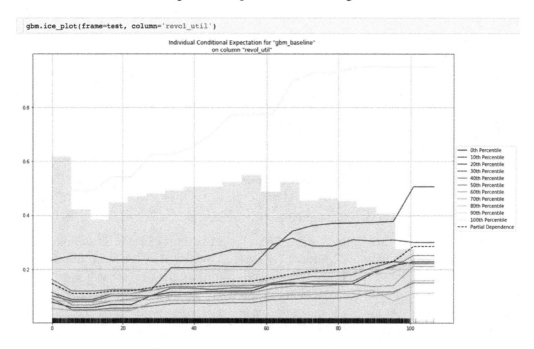

Figure 7.12 – Global ICE plot for revolving utilization

The global ICE plot for a variable is an expansion of the partial dependence plot for that variable. The partial dependence plot displays how the mean response is related to the values of a specific variable. Shading, as shown in *Figure 7.11*, indicates the variability of the partial dependence line. The global ICE plot amplifies this by using multiple lines to represent the population. (In the case of categorical variables, the lines are replaced by points and the shading by bars.)

As shown in *Figure 7.12*, the global ICE plot includes lines for the minimum (0th percentile), the deciles (10th percentile through 90th percentile by 10s), the maximum (100th percentile), and the partial dependence itself. This visually portrays the population much more accurately than partial dependence alone. Percentiles that parallel the partial dependence line correspond to segments of the population for which the partial dependence is a good representation. As is often the case, the behavior of the minimum and maximum of a population may be quite different than the mean behavior described by the partial dependence line. In *Figure 7.12*, there are three lines that are different from the others: the minimum, the maximum, and the 10th percentile.

We next turn our attention to local explanations for single models.

Local explanations for single models

The h2o.explain_row method allows a data scientist to investigate local explanations of a model. While global explanations are used for understanding how the model represents the overall population, local explanations give us the ability to interrogate a model on a per-row basis. This can be especially important in business when rows represent customers, as is the case in our Lending Club analysis.

When predictive models are used to make decisions that impact customers directly (for instance, not approving a loan application or raising a customer's insurance rates), global explanations are not sufficient to satisfy business, legal, or regulatory requirements. This is where local explanations are critical.

The explain_row method returns these local explanations for a specified row_index value. The gbm.explain_row(test, row_index=10) command provides a SHAP explanation plot and multiple ICE plots for columns based on variable importance. As with partial dependence plots, the top_n_features or columns parameters can optionally be provided.

The resulting SHAP explanation plot is shown in *Figure 7.13*:

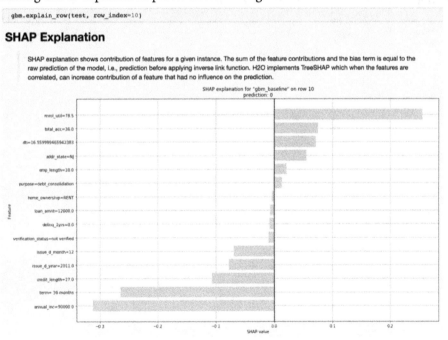

Figure 7.13 – SHAP explanation for index = 10

SHAP explanations show the contributions of each variable to the overall prediction based on Shapley values. For the customer displayed in *Figure 7.13*, the positive SHAP values can be thought of as increasing the probability of loan default, while the negative SHAP values are those decreasing the probability of default. For this customer, the revolving utilization rate of 78.5% is the largest positive contributor to the predicted probability. The largest negative contributor is the annual income of 90,000, which decreases the probability of loan default more than any other variable. SHAP explanations can be used to provide **reason codes** that can help explain the model. Reason codes can also be used as the basis for sharing information directly with the customer, for instance, in adverse action codes that apply to some financial and insurance-related regulatory models.

We next visit some of the ICE plots output from the `explain_row` method.

ICE plots for local explanations

ICE plots are individual or per-row counterparts of partial dependence plots. Just as partial dependence plots for a feature display the mean response of the target variable while varying the feature value, the ICE plot measures the target variable response while varying the feature value for a single row. Consider the ICE plot for the address state displayed in *Figure 7.14* as a result of the `gbm.explain_row` call:

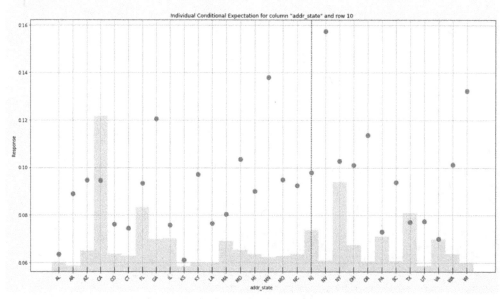

Figure 7.14 – ICE plot for address state

The vertical dark dashed line in *Figure 7.14* represents the actual response for the row in question. In this case, the state is **NJ** (New Jersey) with a response of approximately 0.10. Had the state for this row been **VA** (Virginia), the response would have been lower (about 0.07). Had the state for this row instead been **NV** (Nevada), the response would have been higher, around 0.16.

Consider next *Figure 7.15*, the ICE plot for the term of the loan:

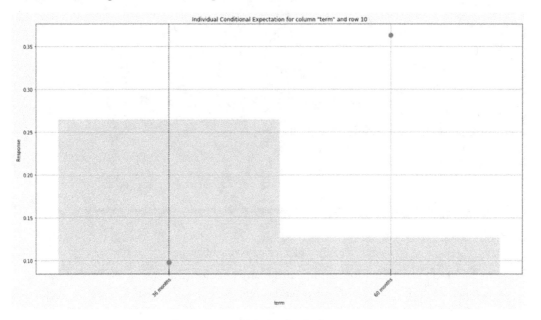

Figure 7.15 – ICE plot for the loan term

The SHAP explanation value in *Figure 7.13* for a term of 36 months was the second-largest negative factor (it reduced the probability of loan default the most after annual income, which was the largest). According to *Figure 7.15*, a loan term of 60 months would have resulted in a default probability of slightly more than 0.35, significantly higher than the approximate 0.10 probability of default with a term of 36 months. While SHAP explanations and ICE plots are measuring two different things, their interpretations can be used jointly to understand the behavior of a particular prediction.

The last ICE plot we consider is for revolving utilization, a numerical rather than categorical feature. This plot is shown in *Figure 7.16*:

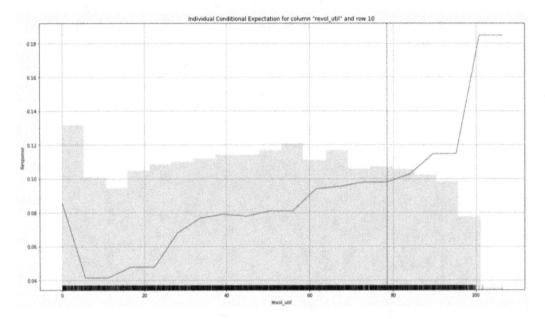

Figure 7.16 – ICE plot for revolving utilization

Revolving utilization was the most significant positive factor (increasing the probability of loan default) according to the SHAP explanations in *Figure 7.13*. The ICE plot in *Figure 7.16* shows the relationship between response and the value of revolving utilization. Had `revol_util` been 50%, the probability of loan default would have been reduced to approximately 0.08. At 20%, the probability of default would be approximately 0.05. If this customer were denied a loan, the high value of revolving utilization would be a defensible reason. Results of the corresponding ICE plot can be used to inform the customer of steps they could take to qualify for the loan.

Global explanations for multiple models

In determining which model to promote into production, for instance, from an AutoML run, the data scientist could rely purely on predictive model metrics. This could mean simply promoting the model with the best AUC value. However, there is a lot of information that could be used to help in this decision, with predictive power being only one of multiple criteria.

The global and local explain features of H2O provide additional information that is useful for evaluating models in conjunction with predictive attributes. We demonstrate it using the `check` AutoML object from *Chapter 5*, *Advanced Model Building – Part 1*.

The code to launch global explanations for multiple models is simply as follows:

```
check.explain(test)
```

This results in a variable importance heatmap, model correlation heatmap, and multiple-model partial dependence plots. We will review each of these in order.

Variable importance heatmap

The variable importance heatmap visually combines the variable importance plots for multiple models by adding color as a dimension to be viewed along with variables (as rows) and models (as columns). The variable importance heatmap produced by check.explain is shown in *Figure 7.17*:

Variable Importance Heatmap

Variable importance heatmap shows variable importance across multiple models. Some models in H2O return variable importance for one-hot (binary indicator) encoded versions of categorical columns (e.g. Deep Learning, XGBoost). In order for the variable importance of categorical columns to be compared across all model types we compute a summarization of the the variable importance across all one-hot encoded features and return a single variable importance for the original categorical feature. By default, the models and variables are ordered by their similarity.

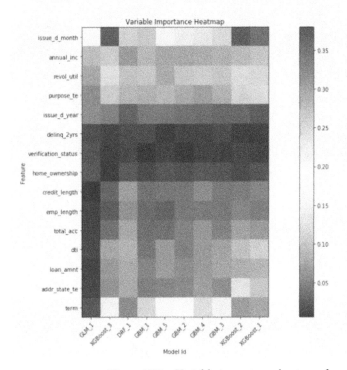

Figure 7.17 – Variable importance heatmap for an AutoML object

Variable importance values are coded as a color continuum from blue (cold) for low values to red (hot) for high values. The resulting figure is visually meaningful. In *Figure 7.17*, vertical bands correspond to each model and horizontal bands correspond to individual features. Vertical bands that are similar indicate a high level of correspondence between how models use their features. For instance, the `XGBoost_1` and `XGBoost_2` models (the last two columns) display similar patterns.

You also see horizontal bands of similar color for variables such as `delinq_2yrs`, `verification_status`, or to a lesser extent, `annual_inc`. This indicates that all the candidate models treat these variables with comparable importance. The `term` variable in the last row is the most visually striking, being heterogeneous across models. These models don't agree on its absolute importance. However, you must be careful not to read too much into this. Notice that for `term`, the *relative* importance is the same for six of the ten models (all but the blue squares: `DRF_1`, `GBM_4`, `XGBoost_2`, and `XGBoost_1`). For these six models, `term` is the most important feature although its exact value varies widely.

The code to create this display directly is as follows:

```
check.varimp_heatmap()
```

Let's next consider the model correlation heatmap.

Model correlation heatmap

The variable importance heatmap allows us to compare multiple models in terms of how they view and use their component variables. The model correlation heatmap addresses a different question: *How correlated are the predictions from these different models?* To answer this, we turn to the model correlation heatmap in *Figure 7.18*:

Model Correlation

This plot shows the correlation between the predictions of the models. For classification, frequency of identical predictions is used. By default, models are ordered by their similarity (as computed by hierarchical clustering). Interpretable models, such as GAM, GLM, and RuleFit are highlighted using red colored text.

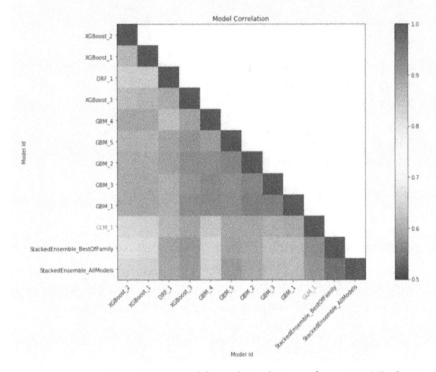

Figure 7.18 – Model correlation heatmap for an AutoML object

The darkest blocks along the diagonal of *Figure 7.18* show a perfect correlation between a model and itself. Sequentially lighter shading describes decreasing correlation between models. How might you use this display to determine which model to promote to production?

This is where business or regulatory constraints can come into play. In our example, `StackedEnsemble_AllModels` had the best model performance in terms of AUC. Suppose that we are not allowed to promote an ensemble model into production, for whatever reason. The single models that are most highly correlated with our best model include `XGBoost_3`, `GBM_5`, and `GLM_1`. These could then become candidates to promote into production, with a final decision based on additional criteria (perhaps the AUC value on the test set).

If one of those additional criteria is native interpretability, then `GLM_1` for this AutoML object is the only choice. Note that interpretable models are indicated with a red-colored font in the model correlation heatmap.

We can create this display directly using the following:

```
check.model_correlation_heatmap(test)
```

Let's move on to introduce partial dependence plots for multiple models in the next subsection.

Multiple-model partial dependence plots

The third output from the `explain` method for multiple models is an extension of the partial dependence plot. For categorical variables, plot symbols and colors corresponding to different models are displayed on an individual plot. *Figure 7.19* is an example using the `term` variable:

Partial Dependence Plots

Partial dependence plot (PDP) gives a graphical depiction of the marginal effect of a variable on the response. The effect of a variable is measured in change in the mean response. PDP assumes independence between the feature for which is the PDP computed and the rest.

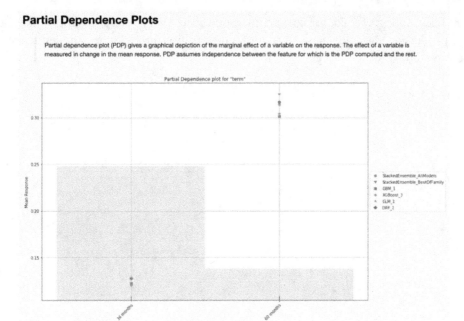

Figure 7.19 – Multiple model partial dependence plot for a loan term

For numeric variables, multiple models are represented by different colored lines on the same partial dependence plot. *Figure 7.20* is an example of this using the `revol_util` variable:

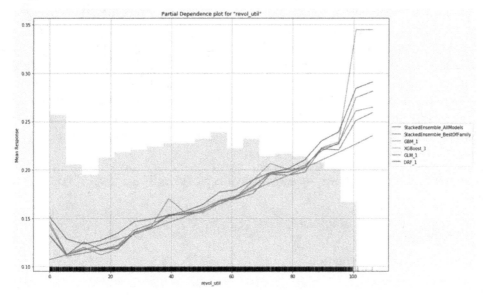

Figure 7.20 – Multiple model partial dependence plot for revolving utilization

In *Figure 7.19* and *Figure 7.20*, the competing models yield very similar results. This is not always the case. For example, *Figure 7.21* shows the multiple model partial dependence plot for annual income:

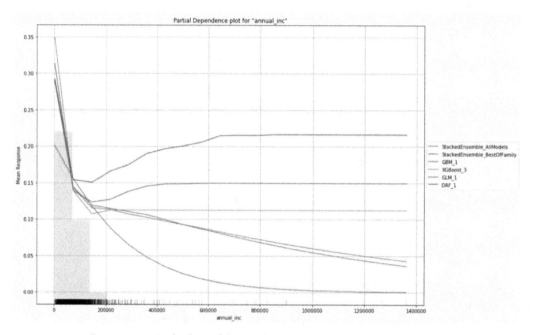

Figure 7.21 – Multiple model partial dependence plot for annual income

Although most of the models in *Figure 7.21* are similar for lower incomes, they diverge rather drastically as incomes increase. This is partially due to the very small sample sizes in the tails of the annual income distribution. The data scientist may also decide to disqualify certain models based on unrealistic or unreasonable tail behavior. For example, based on our experience, it does not make sense for loan default risk to increase as annual income increases. At worst, we would expect no relationship between income and default beyond a certain point. We are more likely to expect a monotonic decrease in loan default as income increases. Based on this reasoning, we would remove the models for the top two lines (DRF_1 and GBM_1) from consideration.

As with other explain methods, we can create this plot directly using the following command:

```
check.pd_multi_plot(test, column='annual_inc')
```

We next visit model documentation.

Automated model documentation (H2O AutoDoc)

One of the important roles a data science team performs in an enterprise setting is documenting the history, attributes, and performance of models that are put into production. At a minimum, model documentation should be part of a data science team's best practices. More commonly in an enterprise setting, thorough model documentation or whitepapers are mandated to satisfy internal and external controls as well as regulatory or compliance requirements.

As a rule, model documentation should be comprehensive enough to allow for the recreation of the model being documented. This entails identifying all data sources, including training and test data characteristics, specifying hardware system components, noting software versions, modeling code, software settings and seeds, modeling assumptions adopted, alternative models considered, performance metrics and appropriate diagnostics, and anything else necessary based on business or regulatory conditions. This process, while vital, is time-consuming and can be tedious.

H2O AutoDoc is a commercial software product that automatically creates comprehensive documentation for models built in H2O-3 and scikit-learn. A similar capability has existed in H2O.ai's **Driverless AI**, a commercial product that combines automatic feature engineering with enhanced AutoML to build and deploy supervised learning models. AutoDoc has been successfully used for documenting models now in production. We present a brief introduction to automatic document creation using AutoDoc here:

1. After a model object has been created, we import the `Config` and `render_autodoc` modules into Python:

   ```
   from h2o_autodoc import Config
   from h2o_autodoc import render_autodoc
   ```

2. Next, we will specify the output file path:

   ```
   config = Config(output_path = "autodoc_report.docx")
   ```

3. Then, we will render the report by passing the configuration information and model object:

   ```
   doc_path = render_autodoc(h2o=h2o, config=config,
                             model=gbm)
   ```

4. Once the report is created, the location of the report can be indicated using the following:

```
print(doc_path)
```

Figure 7.22 shows the table of contents for a 44-page report created by H2O AutoDoc in Microsoft Word:

H2O-3 Experiment

Generated by: AutoDoc Team

Generated on: 2020-11-10 09:54

H2O-3 Experiment 1
- Experiment Overview 1
- Data Overview 3
- Validation Strategy 7
- Feature Importance 7
- Final Model 10
- Alternative Models 29
- Partial Dependence Plots 36
- Model Reproducibility 40
- Appendix 42

Figure 7.22 – Table of contents for model documentation created by H2O AutoDoc

The advantages of thorough documentation produced in a consistent manner with a minimal amount of manual effort are self-evident. Output as either a Microsoft Word document or in markdown format, the reports can be individually edited and further customized. Report templates are also easily edited, allowing a data science team to have a different report structure for different uses: internal whitepaper or report for regulatory review, for example. The AutoDoc capability is consistently one of the best-loved features for H2O software for the enterprise.

Summary

In this chapter, we reviewed multiple model performance metrics and learned how to choose one for evaluating a model's predictive performance. We introduced Shapley values through some simple examples to further understand their purpose and use in predictive model evaluation. Within H2O, we used the `explain` and `explain_row` commands to create global and local explanations for a single model. We learned how to interpret the resulting diagnostics and visualizations to gain trust in a model. For AutoML objects and other lists of models, we generated global and local explanations and saw how to use them alongside model performance metrics to weed out inappropriate candidate models. Putting it all together, we can now evaluate tradeoffs between model performance, scoring speed, and explanations in determining which model to put into production. Finally, we discussed the importance of model documentation and showed how H2O AutoDoc can automatically generate detailed documentation for any model built in H2O (or scikit-learn).

In the next chapter, we will put everything we have learned about building and evaluating models in H2O together to create a deployment-ready model for predicting bad loans in the Lending Club data.

8
Putting It All Together

In this chapter, we will revisit the *Lending Club Loan Application* data that we first introduced in *Chapter 3, Fundamental Workflow – Data to Deployable Model*. This time, we begin the way most data science projects do, that is, with a raw data file and a general objective or question. Along the way, we will refine both the data and the problem statements so that they are relevant to the business and can be answered by the available data. Data scientists rarely begin with modeling-ready data; therefore, the treatment in this chapter more accurately reflects the job of a data scientist in the enterprise. We will then model the data and evaluate various candidate models, updating them as required, until we arrive at a final model. We will evaluate the final model and illustrate the preparation steps required for model deployment. This reinforces what we introduced in *Chapter 5* through *Chapter 7*.

By the end of this chapter, you will be able to take an unstructured problem with a raw data source and create a deployable model to answer a refined predictive question. For completeness, we will include all the code required to do each step of data preparation, feature engineering, model building, and evaluation. In general, any code that has already been covered in *Chapter 5* through *Chapter 7* will be left uncommented.

This chapter is divided into four sections, each of which has individual steps. The sections are listed as follows:

- Data wrangling
- Feature engineering
- Model building and evaluation
- Preparation for model pipeline deployment

Technical requirements

If you still have not set up your H2O environment at this stage, to do so, please see *Appendix – Alternative Methods to Launch H2O Clusters*.

Data wrangling

It is frequently said that 80–90% of a data scientist's job is dealing with data. At a minimum, you should understand the data granularity (that is, what the rows represent) and know what each column in the dataset means. Presented with a raw data source, there are multiple steps required to clean, organize, and transform your data into a modeling-ready dataset format.

The dataset used for the *Lending Club* example in *Chapters 3*, *5*, and *7* was derived from a raw data file that we begin with here. In this section, we will illustrate the following steps:

1. Import the raw data and determine which columns to keep.
2. Define the problem, and create a response variable.
3. Convert the implied numeric data from strings into numeric values.
4. Clean up any messy categorical columns.

Let's begin with the first step: importing the data.

Importing the raw data

We import the raw data file using the following code:

```
input_csv = "rawloans.csv"
loans = h2o.import_file(input_csv,
            col_types = {"int_rate": "string",
                         "revol_util": "string",
```

```
            "emp_length": "string",
            "verification_status": "string"})
```

A dictionary in the `h2o.import_file` code specifies the input column type of `string` for four of the input variables: `int_rate`, `revol_util`, `emp_length`, and `verification_status`. Specifying the column type explicitly ensures that the column is read in as the modeler intended. Without this code, these string variables might have been read as categorical columns with multiple levels.

The dataset dimensions are obtained by the following command:

```
loans.dim
```

This returns 42,536 rows (corresponding to 42,536 customer credit applications) and 52 columns. Next, we specify the 22 columns we wish to keep for our analysis:

```
keep = ['addr_state', 'annual_inc', 'delinq_2yrs',
        'dti', 'earliest_cr_line', 'emp_length', 'grade',
        'home_ownership', 'inq_last_6mths', 'installment',
        'issue_d', 'loan_amnt', 'loan_status',
        'mths_since_last_delinq', 'open_acc', 'pub_rec',
        'purpose', 'revol_bal', 'revol_util', 'term',
        'total_acc', 'verification_status']
```

And we want to remove the remaining columns using the `drop` method:

```
remove = list(set(loans.columns) - set(keep))
loans = loans.drop(remove)
```

But what about the 30 columns that we removed? They contained things such as text descriptions of the purpose of the loan, additional customer information such as the address or zip code, columns with almost completely missing information or other data quality issues, and more. Selecting the appropriate columns from a raw data source is an important task that takes much time and effort on the part of the data scientist.

The columns we keep are those we believe are most likely to be predictive. Explanations for each column are listed as follows:

- `addr_state`: This is the US state where the borrower resides.
- `annual_inc`: This is the self-reported annual income of the borrower.
- `delinq_2yrs`: This is the number of times the borrower has been more than 30 days late in payments during the last 2 years.

- `dti`: This is the debt-to-income ratio (current debt divided by income).
- `earliest_cr_line`: This is the date of the earliest credit line (generally, longer credit histories correlate with better credit risk).
- `emp_length`: This is the length of employment.
- `grade`: This is a risk rating from A to G assigned to the loan by the lender.
- `home_ownership`: Does the borrower own a home or rent?
- `inq_last_6mths`: This is the number of credit inquiries in the last 6 months.
- `installment`: This is the monthly amount owed by the borrower.
- `issue_d`: This is the date the loan was issued.
- `loan_amnt`: This is the total amount lent to the borrower.
- `loan_status`: This is a category.
- `mths_since_last_delinq`: This is the number of months since the last delinquency.
- `open_acc`: This is the number of open credit lines.
- `pub_rec`: This is the number of derogatory public records (bankruptcies, tax liens, and judgments).
- `purpose`: This is the borrower's stated purpose for the loan.
- `revol_bal`: This is the revolving balance (that is, the amount owed on credit cards at the end of the billing cycle).
- `revol_util`: This is the revolving utilization (that is, the amount of credit used divided by the total credit available to the borrower).
- `term`: This is the number of payments on the loan in months (either 36 or 60).
- `total_acct`: This is the borrower's total number of credit lines.
- `verification_status`: This tells us whether the income was verified or not.

Assuming our data columns have been properly selected, we can move on to the next step: creating the response variable.

Defining the problem and creating the response variable

The creation of the response variable depends on the problem definition. The goal of this use case is to predict which customers will default on a loan. A model that predicts a loan default needs a response variable that differentiates between good and bad loans. Let's start by investigating the `loan_status` variable using the following code:

```
loans["loan_status"].table().head(20)
```

This produces a table with all possible values of the loan status stored in our data:

```
loans["loan_status"].table().head(20)
```

loan_status	Count
Charged Off	5435
Current	3351
Default	7
Does not meet the credit policy. Status:Charged Off	761
Does not meet the credit policy. Status:Current	53
Does not meet the credit policy. Status:Fully Paid	1933
Does not meet the credit policy. Status:In Grace Period	2
Fully Paid	30843
In Grace Period	60
Late (16-30 days)	16
Late (31-120 days)	74

Figure 8.1 – Loan status categories from the raw Lending Club loan default dataset

As you can see in *Figure 8.1*, the `loan_status` variable is relatively complex, containing 11 categories that are somewhat redundant or overlapping. For instance, `Charged Off` indicates that 5,435 loans were bad. `Default` contains another 7. `Fully Paid` shows that 30,843 loans were good. Some loans, for example, those indicated by the `Current` or `Late` categories, are still ongoing and so are not yet good or bad.

Multiple loans were provided that did not meet the credit policy. Why this was allowed is unclear and worth checking with the data source. Did the credit policy change so that these loans are of an earlier vintage? Are these formal overrides or were they accidental? Whatever the case might be, these categories hint at a different underlying population that might require our attention. Should we remove these loans altogether, ignore the issue by collapsing them into their corresponding categories, or create a `Meets Credit Policy` indicator variable and model them directly? A better understanding of the data would allow the data scientist to make an informed decision.

In the end, we need a binary response variable based on a population of loans that have either been paid off or defaulted. First, filter out any ongoing loans.

Removing ongoing loans

We need to build our model with only those loans that have either defaulted or been fully paid off. Ongoing loans have `loan_status` such as `Current` or `In Grace Period`. The following code captures the rows whose statuses indicate ongoing loans:

```
ongoing_status = [
    "Current",
    "In Grace Period",
    "Late (16-30 days)",
    "Late (31-120 days)",
    "Does not meet the credit policy.  Status:Current",
    "Does not meet the credit policy.  Status:In Grace Period"
]
```

We use the following code to remove those ongoing loans and display the status for the remaining loans:

```
loans = loans[~loans["loan_status"].isin(ongoing_status)]
loans["loan_status"].table()
```

The resulting status categories are shown in *Figure 8.2*:

```
loans["loan_status"].table()
```

loan_status	Count
Charged Off	5435
Default	7
Does not meet the credit policy. Status:Charged Off	761
Does not meet the credit policy. Status:Fully Paid	1933
Fully Paid	30843

Figure 8.2 – Loan status categories after filtering the ongoing loans

Note that, in *Figure 8.2*, five categories of loan status now need to be summarized in a binary response variable. This is detailed in the next step.

Defining the binary response variable

We start by forming a `fully_paid` list to summarize the loan status categories:

```
fully_paid = [
    "Fully Paid",
    "Does not meet the credit policy.  Status:Fully Paid"
]
```

Next, let's create a binary response column, `bad_loan`, as an indicator for any loans that were not completely paid off:

```
response = "bad_loan"
loans[response] = ~(loans["loan_status"].isin(fully_paid))
loans[response] = loans[response].asfactor()
```

Finally, remove the original loan status column:

```
loans = loans.drop("loan_status")
```

We remove the original loan status column because the information we need for building our predictive model is now contained in the `bad_loan` response variable.

Next, we will convert string data into numeric values.

Converting implied numeric data from strings into numeric values

There are various ways that data can be messy. In the preceding step, we saw how variables can sometimes contain redundant categories that might benefit from summarization. The format in which data values are displayed and stored can also cause problems. Therefore, the 28% that we naturally interpret as a number is, typically, input as a character string by a computer. Converting implied numeric data into actual numeric data is a very typical data quality task.

Consider the `revol_util` and `emp_length` columns:

```
loans[["revol_util", "emp_length"]].head()
```

The output is shown in the following screenshot:

```
loans[["revol_util", "emp_length"]].head()
```

revol_util	emp_length
83.7%	10+ years
9.4%	< 1 year
98.5%	10+ years
21%	10+ years
28.3%	3 years
87.5%	9 years
32.6%	4 years
36.5%	< 1 year
20.6%	5 years
67.1%	10+ years

Figure 8.3 – Variables stored as strings to be converted into numeric values

The `revol_util` variable, as shown in *Figure 8.3*, is inherently numeric but has a trailing percent sign. In this case, the solution is simple: strip the % sign and convert the strings into numeric values. This can be done with the following code:

```
x = "revol_util"
loans[x] = loans[x].gsub(pattern="%", replacement="")
loans[x] = loans[x].trim()
loans[x] = loans[x].asnumeric()
```

The `gsub` method substitutes `%` with a blank space. The `trim` method removes any whitespace in the string. The `asnumeric` method converts the string value into a number.

The `emp_length` column is only slightly more complex. First, we need to strip out the year or years term. Also, we must deal with the < and + signs. If we define < 1 as 0 and 10+ as 10, then `emp_length` can also be cast as numeric. This can be done using the following code:

```
x = "emp_length"
loans[x] = loans[x].gsub(pattern="([ ]*+[a-zA-Z].*)|(n/a)",
                         replacement="")
loans[x] = loans[x].trim()
loans[x] = loans[x].gsub(pattern="< 1", replacement="0")
loans[x] = loans[x].gsub(pattern="10\\+", replacement="10")
loans[x] = loans[x].asnumeric()
```

Next, we will complete our data wrangling steps by cleaning up any messy categorical columns.

Cleaning up messy categorical columns

The last step in preparation for feature engineering and modeling is clarifying the options or levels in often messy categorical columns. This standardization task is illustrated by the `verification_status` variable. Use the following code to find the levels of `verification_status`:

```
loans["verification_status"].head()
```

The results are displayed in *Figure 8.4*:

```
loans["verification_status"].head()
```

verification_status

VERIFIED - income

VERIFIED - income source

not verified

VERIFIED - income source

VERIFIED - income source

VERIFIED - income source

VERIFIED - income source

VERIFIED - income

not verified

VERIFIED - income source

Figure 8.4 – The categories of the verification status from the raw data

Because there are multiple values in *Figure 8.4* that mean verified (VERIFIED - income and VERIFIED - income source), we simply replace them with verified. The following code uses the sub method for easy replacement:

```
x = "verification_status"
loans[x] = loans[x].sub(pattern = "VERIFIED - income source",
                        replacement = "verified")
loans[x] = loans[x].sub(pattern = "VERIFIED - income",
                        replacement = "verified")
loans[x] = loans[x].asfactor()
```

After completing all our data wrangling steps, we will move on to feature engineering.

Feature engineering

In *Chapter 5*, *Advanced Model Building – Part I*, we introduced some feature engineering concepts and discussed target encoding at length. In this section, we will delve into feature engineering in a bit more depth. We can organize feature engineering as follows:

- Algebraic transformations

- Features engineered from dates

- Simplifying categorical variables by combining categories

- Missing value indicator functions

- Target encoding categorical columns

The ordering of these transformations is not important except for the last one. Target encoding is the only transformation that requires data to be split into train and test sets. By saving it for the end, we can apply the other transformations to the entire dataset at once rather than separately to the training and test splits. Also, we introduce stratified sampling for splitting data in H2O-3. This has very little impact on our current use case but is important when data is highly imbalanced, such as in fraud modeling.

In the following sections, we include all our feature engineering code for completeness. Code that has been introduced earlier will be merely referenced, while new feature engineering tasks will merit discussion. Let's begin with algebraic transformations.

Algebraic transformations

The most straightforward form of feature engineering entails taking simple transformations of the raw data columns: the log, the square, the square root, the differences in columns, the ratios of the columns, and more. Often, the inspiration for these transformations comes from an underlying theory or is based on subject-matter expertise.

The `credit_length` variable, as defined in *Chapter 5*, *Advanced Model Building – Part I*, is one such transformation. Recall that this is created with the following code:

```
loans["credit_length"] = loans["issue_d"].year() - \
    loans["earliest_cr_line"].year()
```

The justification for this variable is based on a business observation: customers with longer credit histories tend to be at lower risk of defaulting. Also, we drop the `earliest_cr_line` variable, which is no longer needed:

```
loans = loans.drop(["earliest_cr_line"])
```

Another simple feature we could try is *(annual income)/(number of credit lines)*, taking the log for distributional and numerical stability. Let's name it `log_inc_per_acct`. This ratio makes intuitive sense: larger incomes should be able to support a greater number of credit lines. This is similar to the debt-to-income ratio in intent but captures slightly different information. We can code it as follows:

```
x = "log_inc_per_acct"
loans[x] = loans['annual_inc'].log() - \
    loans['total_acc'].log()
```

Next, we will consider the second feature engineering task: encoding information from dates.

Features engineered from dates

As noted in *Chapter 5*, *Advanced Model Building – Part I*, there is a wealth of information contained in date values that are potentially predictive. To the `issue_d_year` and `issue_d_month` features that we created earlier, we add `issue_d_dayOfWeek` and `issue_d_weekend` as new factors. The code to do this is as follows:

```
x = "issue_d"
loans[x + "_year"] = loans[x].year()
loans[x + "_month"] = loans[x].month().asfactor()
loans[x + "_dayOfWeek"] = loans[x].dayOfWeek().asfactor()
weekend = ["Sat", "Sun"]
loans[x + "_weekend"] = loans[x + "_dayOfWeek"].isin(weekend)
loans[x + "_weekend"] = loans[x + "_weekend"].asfactor()
```

At the end, we drop the original date variable:

```
loans = loans.drop(x)
```

Next, we will address how to simplify categorical variables at the feature engineering stage.

Simplifying categorical variables by combining categories

In the data wrangling stage, we cleaned up the messy categorical levels for the `verification_status` column, removing redundant or overlapping level definitions and making categories mutually exclusive. On the other hand, during this feature engineering stage, the category levels are already non-overlapping and carefully defined. The data values themselves, for instance, small counts for certain categories, might suggest some engineering approaches to improve predictive modeling.

Summarize the `home_ownership` categorical variable using the following code:

```
x = "home_ownership"
loans[x].table()
```

The tabled results are shown in the following screenshot:

```
x = "home_ownership"
loans[x].table()
```

home_ownership	Count
MORTGAGE	17056
NONE	8
OTHER	135
OWN	2997
RENT	18783

Figure 8.5 – Levels of the raw home ownership variable

In *Figure 8.5*, although there are five recorded categories within home ownership, the largest three have thousands of observations: MORTGAGE, OWN, and RENT. The remaining two, NONE and OTHER, are so infrequent (8 and 135, respectively) that we will combine them with OWN to create an expanded OTHER category.

> **Collapsing Data Categories**
>
> Depending on the inference we want to make, or our understanding of the problem, it might make more sense to collapse NONE and OTHER into the RENT or MORTGAGE categories.

The procedure for combining the categorical levels is shown by the following command:

```
loans[x].levels()
```

This is given by replacing the NONE and OWN level descriptions with OTHER and assigning it to a new variable, home_3cat, as shown in the following code:

```
lvls = ["MORTGAGE", "OTHER", "OTHER", "OTHER", "RENT"]
loans["home_3cat"] = \
    loans[x].set_levels(lvls).ascharacter().asfactor()
```

Then, we drop the original home_ownership column:

```
loans = loans.drop(x)
```

Next, we will visit how to create useful indicator functions for missing data.

Missing value indicator functions

When data is not missing at random, the pattern of missingness might be a source of predictive information. In other words, sometimes, the fact that a value is missing is as, or more, important than the actual value itself. Especially in cases where missing values are abundant, creating a missing value indicator function can prove helpful.

The most interesting characteristic of employment length, emp_length, is whether the value for a customer is missing. Simple pivot tables show that the proportion of bad loans is 26.3% for customers with missing emp_length values and 18.0% for non-missing values. That disparity in default rates suggests using a missing value indicator function as a predictor.

The code for creating a missing indicator function for the emp_length variable is simple:

```
loans["emp_length_missing"] = loans["emp_length"] == None
```

Here, the new emp_length_missing column contains the indicator function. Unlike the other features that we engineered earlier, the original emp_length column does not need to be dropped as a possible predictor.

Next, we will turn to target encoding categorical columns.

Target encoding categorical columns

In *Chapter 5*, *Advanced Model Building – Part I*, we introduced target encoding in H2O-3 in some detail. As a prerequisite to target encoding, recall that a train and test set was required. We split the data using the `split_frame` method with code similar to the following:

```
train, test = loans.split_frame(ratios = [0.8], seed = 12345)
```

The `split_frame` method creates a completely random sample split. This approach is required for all regression models and works well for relatively balanced classification problems. However, when binary classification is highly imbalanced, stratified sampling should be used instead.

Stratified sampling for binary classification data splits

Stratified sampling for binary classification works by separately sampling the good and bad loans. In other words, recall that 16% of the loans in our Lending Club dataset are bad. We wish to split the data into 80% train and 20% test datasets. If we separately sample 20% of the bad loans and 20% of the good loans and then combine them, we have a test dataset that preserves the 16% bad loan percentage. Combining the remaining data results in a 16% bad loan percentage in our training data. Therefore, stratified sampling preserves the original category ratios.

We use the `stratified_split` method on the response column to create a new variable named `split`, which contains the `train` and `test` values, as shown in the following code:

```
loans["split"] = loans[response].stratified_split(\
    test_frac = 0.2, seed = 12345)
loans[[response,"split"]].table()
```

The results of the stratified split are shown in the following screenshot:

```
loans["split"] = loans[response].stratified_split(\
    test_frac = 0.2, seed = 12345)
loans[[response,"split"]].table()
```

bad_loan	split	Counts
0	train	26221
0	test	6555
1	train	4963
1	test	1241

Figure 8.6 – The stratified split of loan data into train and test

We use the `split` column to create a Boolean mask for deriving the `train` and `test` datasets, as shown in the following code:

```
mask = loans["split"] == "train"
train = loans[mask, :].drop("split")
test = loans[~mask, :].drop("split")
```

Note that we drop the `split` column from both datasets after their creation. Now we are ready to target encode using these train and test splits.

Target encoding the Lending Club data

The following code for target encoding the `purpose` and `addr_state` variables is similar to the code from *Chapter 5, Advanced Model Building – Part I*, which we have included here without discussion:

```
from h2o.estimators import H2OTargetEncoderEstimator
encoded_columns = ["purpose", "addr_state"]
train["fold"] = train.kfold_column(n_folds = 5, seed = 25)
te = H2OTargetEncoderEstimator(
    data_leakage_handling = "k_fold",
    fold_column = "fold",
    noise = 0.05,
    blending = True,
    inflection_point = 10,
    smoothing = 20)
te.train(x = encoded_columns,
        y = response,
        training_frame = train)
train_te = te.transform(frame = train)
test_te = te.transform(frame = test, noise = 0.0)
```

Next, we redefine the `train` and `test` datasets, dropping the encoded columns from the target-encoded `train_te` and `test_te` splits. Also, we also drop the `fold` column from the `train_te` dataset (note that it does not exist in the `test_te` dataset). The code is as follows:

```
train = train_te.drop(encoded_columns).drop("fold")
test = test_te.drop(encoded_columns)
```

With our updated `train` and `test` datasets, we are ready to tackle the model building and evaluation processes.

Model building and evaluation

Our approach to model building starts with AutoML. Global explainability applied to the AutoML leaderboard either results in picking a candidate model or yields insights that we feed back into a new round of modified AutoML models. This process can be repeated if improvements in modeling or explainability are apparent. If a single model rather than a stacked ensemble is chosen, we can show how an additional random grid search could produce better models. Then, the final candidate model is evaluated.

The beauty of this approach in H2O-3 is that the modeling heavy lifting is done for us automatically with AutoML. Iterating through this process is straightforward, and the improvement cycle can be repeated, as needed, until we have arrived at a satisfactory final model.

We organize the modeling steps as follows:

1. Model search and optimization with AutoML.
2. Investigate global explainability with the AutoML leaderboard models.
3. Select a model from the AutoML candidates, with an optional additional grid search.
4. Final model evaluation.

Model search and optimization with AutoML

The model build process using H2O-3 AutoML was extensively introduced in *Chapter 5, Advanced Model Building – Part I*. Here, we will follow a virtually identical process to create a leaderboard of models fit by AutoML. For clarity, we redefine our `response` column and `predictors` before removing the `bad_loan` response from the set of `predictors`:

```
response = "bad_loan"
predictors = train.columns
predictors.remove(response)
```

Our AutoML parameters only exclude deep learning models, allowing the process to run for up to 30 minutes, as shown in the following code snippet:

```
from h2o.automl import H2OAutoML
aml = H2OAutoML(max_runtime_secs = 1800,
```

```
                    exclude_algos = ['DeepLearning'],
                    seed = 12345)
aml.train(x = predictors,
          y = response,
          training_frame = train)
```

As demonstrated in *Chapter 5*, *Advanced Model Building – Part I*, we can access H2O Flow to monitor the model build process in more detail. Once the training on the `aml` object finishes, we proceed to investigate the resulting models with global explainability.

Investigating global explainability with AutoML models

In *Chapter 7*, *Understanding ML Models*, we outlined the use of global explainability for a series of models produced by AutoML. Here, we will follow the same procedure by calling the `explain` method with the `test` data split:

```
aml.explain(test)
```

The resulting AutoML leaderboard is shown in the following screenshot:

Leaderboard

Leaderboard shows models with their metrics. When provided with H2OAutoML object, the leaderboard shows 5-fold cross-validated metrics by default (depending on the H2OAutoML settings), otherwise it shows metrics computed on the frame. At most 20 models are shown by default.

model_id	auc	logloss	aucpr	mean_per_class_error	rmse	mse
StackedEnsemble_AllModels_AutoML_20211127_054659	0.735817	0.388986	0.34871	0.334755	0.346336	0.119948
StackedEnsemble_BestOfFamily_AutoML_20211127_054659	0.734067	0.389716	0.346406	0.331813	0.346569	0.12011
XGBoost_grid__1_AutoML_20211127_054659_model_6	0.729386	0.391866	0.341858	0.340299	0.34747	0.120735
XGBoost_grid__1_AutoML_20211127_054659_model_1	0.727624	0.393091	0.339654	0.337556	0.348039	0.121131
XGBoost_grid__1_AutoML_20211127_054659_model_7	0.727553	0.392948	0.33825	0.338682	0.348171	0.121223
XGBoost_grid__1_AutoML_20211127_054659_model_8	0.727244	0.393432	0.336407	0.347118	0.348207	0.121248
XGBoost_grid__1_AutoML_20211127_054659_model_9	0.72627	0.393192	0.339535	0.338083	0.348112	0.121182
GBM_grid__1_AutoML_20211127_054659_model_2	0.725916	0.394446	0.331936	0.342641	0.348608	0.121527
GLM_1_AutoML_20211127_054659	0.72459	0.394225	0.333907	0.34216	0.348267	0.12129
XGBoost_grid__1_AutoML_20211127_054659_model_2	0.724087	0.394942	0.330171	0.341693	0.348909	0.121737

Figure 8.7 – The top 10 models of the AutoML leaderboard

The stacked ensemble `AllModels` and `BestOfFamily` models claim the top two positions on the leaderboard in *Figure 8.7*. The best single model is enclosed by a green box and labeled `model_6` from `XGBoost_grid__1`. We will investigate this model a bit further as a possible candidate model.

The **Model Correlation** plot is shown in *Figure 8.8*. The green box indicates the correlation between our candidate XGBoost model and the two stacked ensembles. It confirms that the candidate model has among the highest correlation with the ensembles:

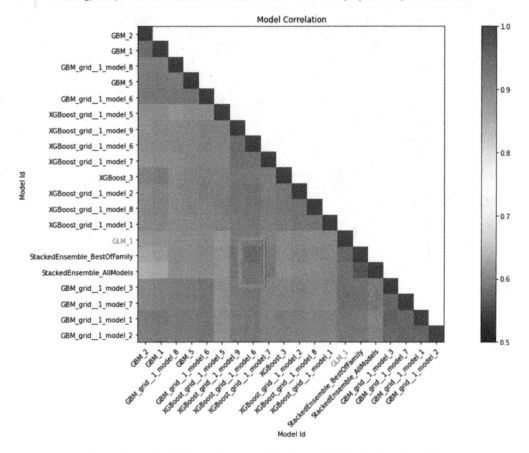

Figure 8.8 – Model Correlation plot for the AutoML leaderboard models

The **Variable Importance Heatmap** diagram in *Figure 8.9* tells us more about the stability of the individual features than about the relationship between the models. The GBM grid models of 1, 2, 3, and 7 cluster together, and the XGBoost grid models of 6, 7, and 9 appear very similar in terms of how important variables are in these models:

Variable Importance Heatmap

Variable importance heatmap shows variable importance across multiple models. Some models in H2O return variable importance for one-hot (binary indicator) encoded versions of categorical columns (e.g. Deep Learning, XGBoost). In order for the variable importance of categorical columns to be compared across all model types we compute a summarization of the the variable importance across all one-hot encoded features and return a single variable importance for the original categorical feature. By default, the models and variables are ordered by their similarity.

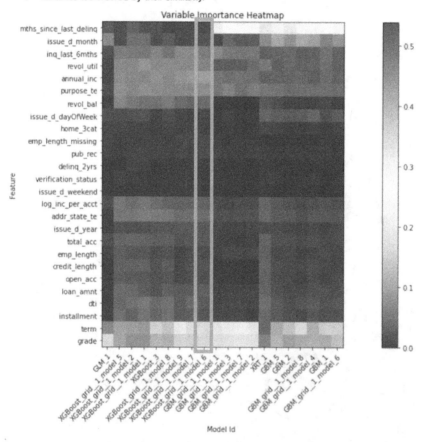

Figure 8.9 – Variable Importance Heatmap for AutoML models

The multiple model **Partial Dependence Plots** (**PDPs**), in conjunction with the variable importance heatmap, yield some valuable insights. *Figure 8.10* shows the PDP for grade, a feature with values from A to G that appear to be increasing at default risk. In other words, the average response for A is less than that for B, which is itself less than that for C, and so forth. This diagnostic appears to be confirming a business rating practice:

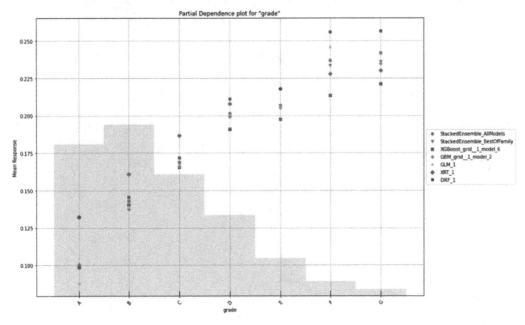

Figure 8.10 – The multiple model PDP for grade

In *Figure 8.11*, the PDP for annual income acts as a diagnostic. Intuitively, an increase in annual income should correspond to a decrease in bad loan rates. We can formally enforce (rather than just hope for) a monotonic decreasing relationship between the annual income and the default rate by adding monotonicity constraints to our model build code:

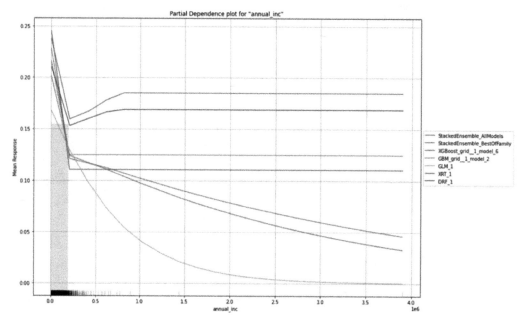

Figure 8.11 – The multiple model PDP for annual income

Monotonicity constraints can be applied to one or more numeric variables in the GBM, XGBoost, and AutoML models in H2O-3. To do so, supply a `monotone_constraints` parameter with a dictionary of variable names and the direction of the monotonicity: 1 for a monotonic increasing relationship and -1 for monotonic decreasing. The following code shows how we add a monotonic decreasing `annual_inc` constraint:

```
maml = H2OAutoML(
        max_runtime_secs = 1800,
        exclude_algos = ['DeepLearning'],
        monotone_constraints = {"annual_inc": -1},
        seed = 12345)
```

> **Monotonic Increasing and Decreasing Constraints**
>
> Formally, the monotonic increasing constraint is a monotonic non-decreasing constraint, meaning that the function must either be increasing or flat. Likewise, the monotonic decreasing constraint is more correctly termed a monotonic non-increasing constraint.

Fitting a constrained model proceeds as usual:

```
maml.train(x = predictors,
           y = response,
           training_frame = train)
```

Here is the `explain` method:

```
maml.explain(test)
```

This produces the leaderboard, as shown in the following screenshot:

Leaderboard

Leaderboard shows models with their metrics. When provided with H2OAutoML object, the leaderboard shows 5-fold cross-validated metrics by default (depending on the H2OAutoML settings), otherwise it shows metrics computed on the frame. At most 20 models are shown by default.

model_id	auc	logloss	aucpr	mean_per_class_error	rmse	mse
StackedEnsemble_AllModels_AutoML_20211127_060117	0.735584	0.388901	0.350368	0.337067	0.346222	0.11987
StackedEnsemble_Monotonic_AutoML_20211127_060117	0.73478	0.389458	0.349005	0.338251	0.346435	0.120017
StackedEnsemble_BestOfFamily_AutoML_20211127_060117	0.734041	0.38967	0.346638	0.334018	0.346535	0.120086
XGBoost_grid__1_AutoML_20211127_060117_model_6	0.729576	0.391748	0.342034	0.340917	0.347398	0.120685
XGBoost_grid__1_AutoML_20211127_060117_model_7	0.727395	0.392795	0.341516	0.342497	0.347851	0.121
XGBoost_grid__1_AutoML_20211127_060117_model_9	0.726913	0.392621	0.341578	0.342318	0.347801	0.120965
XGBoost_grid__1_AutoML_20211127_060117_model_8	0.726476	0.393647	0.335496	0.348137	0.348377	0.121367
XGBoost_grid__1_AutoML_20211127_060117_model_1	0.7262	0.39377	0.337408	0.341047	0.348282	0.1213
XGBoost_grid__1_AutoML_20211127_060117_model_2	0.725496	0.393931	0.334747	0.34883	0.348395	0.121379
GBM_grid__1_AutoML_20211127_060117_model_2	0.725482	0.394689	0.330472	0.339647	0.348709	0.121598

Figure 8.12 – The leaderboard for AutoML with monotonic constraints

The first 10 models of the updated AutoML leaderboard are shown in *Figure 8.12*. Note that a new model has been added, the monotonic stacked ensemble (boxed in red). This stacked ensemble uses, as constituent models, only those that are monotonic. In our case, this means that any DRF and XRT random forest models fit by AutoML would be excluded. Also note that the monotonic version of XGBoost model 6 is once more the leading single model, boxed in green.

Figure 8.13 shows the monotonic multiple model PDP for annual income:

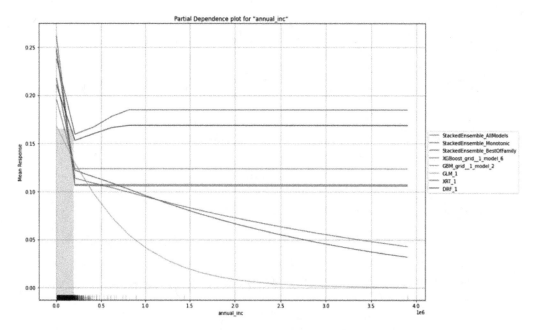

Figure 8.13 – The multiple model PDP for annual income

Note that only two of the models included in the PDP of *Figure 8.13* are not monotonic: the DRF and XRT models. They are both versions of random forest that do not have monotonic options. This plot confirms that the monotonic constraints on annual income worked as intended. (Note that the PDP in *Figure 8.11* is very similar. The models there might have displayed monotonicity, but it was not enforced.)

Next, we will consider how to choose a model from the AutoML leaderboard.

Selecting a model from the AutoML candidates

Once AutoML has created a class of models, it is left to the data scientist to determine which model to put into production. If pure predictive accuracy is the only requirement, then the choice is rather simple: select the top model in the leaderboard (usually, this is the **All Models** stacked ensemble). In the case where monotonic constraints are required, the monotonic stacked ensemble is usually the most predictive.

If business or regulatory constraints only allow a single model to be deployed, then we can select one based on a combination of predictive performance and other considerations, such as the modeling type. Let's select XGBoost model 6 as our candidate model:

```
candidate = h2o.get_model(maml.leaderboard[3, 'model_id'])
```

H2O-3 AutoML does a tremendous job at building and tuning models across multiple modeling types. For individual models, it is sometimes possible to get an improvement in performance via an additional random grid search. We will explore this in the next section.

Random grid search to improve the selected model (optional)

We use the parameters of the candidate model as starting points for our random grid search. The idea is to search within the neighborhood of the candidate model for models that perform slightly better, noting that any improvements found will likely be minor. The stacked ensemble models give us a ceiling for how well an individual model can perform. The data scientist must judge whether the difference between candidate model performance and stacked ensemble performance warrants the extra effort in searching for possibly better models.

We can list the model parameters using the following code:

```
candidate.actual_params
```

Start by importing H2OGridSearch and the candidate model estimator; in our case, that is H2OXGBoostEstimator:

```
from h2o.grid.grid_search import H2OGridSearch
from h2o.estimators import H2OXGBoostEstimator
```

The hyperparameters are selected by looking at the candidate model's actual parameters and searching within the neighborhood of those values. For instance, the sample rate for the candidate model was reported as 80%, and in our hyperparameter tuning, we select a range between 60% and 100%. Likewise, a 60% column sample rate leads us to implement a range between 40% and 80% for the grid search. The hyperparameter tuning code is as follows:

```
hyperparams_tune = {
    'max_depth' : list(range(2, 6, 1)),
    'sample_rate' : [x/100. for x in range(60,101)],
    'col_sample_rate' : [x/100. for x in range(40,80)],
    'col_sample_rate_per_tree': [x/100. for x in
        range(80,101)],
    'learn_rate' : [x/100. for x in range(5,31)]
}
```

We limit the overall runtime of the random grid search to 30 minutes, as follows:

```
search_criteria_tune = {
    'strategy' : "RandomDiscrete",
    'max_runtime_secs' : 1800,
    'stopping_rounds' : 5,
    'stopping_metric' : "AUC",
    'stopping_tolerance': 5e-4
}
```

We add the monotonic constraints to the model and define our grid search:

```
monotone_xgb_grid = H2OXGBoostEstimator(
    ntrees = 90,
    nfolds = 5,
    score_tree_interval = 10,
    monotone_constraints = {"annual_inc": -1},
    seed = 25)
monotone_grid = H2OGridSearch(
    monotone_xgb_grid,
    hyper_params = hyperparams_tune,
    grid_id = 'monotone_grid',
    search_criteria = search_criteria_tune)
```

Then, we train the model:

```
monotone_grid.train(
    x = predictors,
    y = response,
    training_frame = train)
```

Returning to our results after this long training period, we extract the top two models to compare them with our initial candidate model. Note that we order by `logloss`:

```
monotone_sorted = monotone_grid.get_grid(sort_by = 'logloss',
                                          decreasing = False)
best1 = monotone_sorted.models[0]
best2 = monotone_sorted.models[1]
```

Determine the performance of each of these models on the test data split:

```
candidate.model_performance(test).logloss()
best1.model_performance(test).logloss()
best2.model_performance(test).logloss()
```

On the test sample, the logloss for the `candidate` model is 0.3951, `best1` is 0.3945, and `best2` is 0.3937. Based on this criterion alone, the `best2` model is our updated candidate model. The next step is the evaluation of this final model.

Final model evaluation

Having selected `best2` as our final candidate, next, we evaluate this individual model using the `explain` method:

```
final = best2
final.explain(test)
```

We will use the variable importance plot in *Figure 8.14* in conjunction with individual PDPs to understand the impact of the input variables on this model:

Variable Importance

The variable importance plot shows the relative importance of the most important variables in the model.

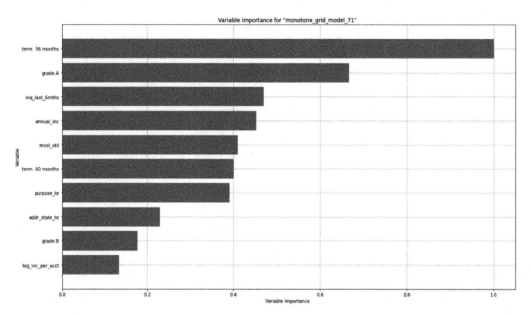

Figure 8.14 – Variable importance for the final model

The `term` variable is by far the most important variable in the final model. Inspecting the PDP for "term" in *Figure 8.15* explains why.

Partial Dependence Plots

Partial dependence plot (PDP) gives a graphical depiction of the marginal effect of a variable on the response. The effect of a variable is measured in change in the mean response. PDP assumes independence between the feature for which is the PDP computed and the rest.

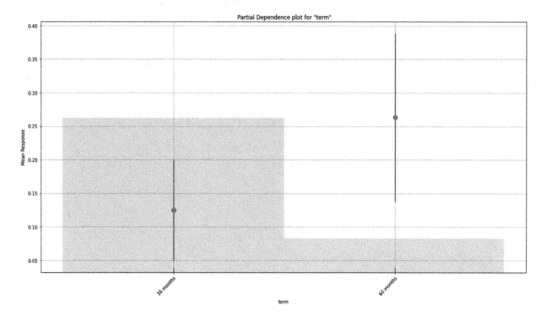

Figure 8.15 – PDP for term

Loans with a term of 36 months have a default rate of around 12%, while 60-month loans have a default rate that jumps to over 25%. Note that because this is an XGBoost model, `term` was parameterized as `term 36 months`.

The next variable in importance is `grade` A. This is an indicator function for one level of the categorical `grade` variable. Looking at the PDP for `grade` in *Figure 8.16*, loans with a level of A only have a 10% default rate with an approximate 5% jump for the next lowest risk grade, B:

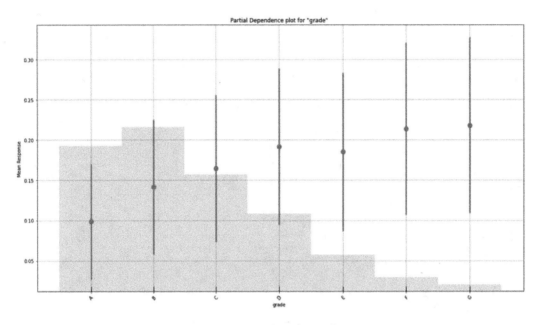

Figure 8.16 – PDP for grade

The next two variables are numeric and roughly equivalent in importance: credit inquiries in the last 6 months (`inq_last_6mths`) and annual income. Their PDPs are shown in *Figures 8.17* and *Figure 8.18*, respectively. The credit inquiries PDP appears to be monotonic except for the right-hand tail. This is likely due to thin data in this upper region of high numbers of inquiries. It would probably make sense to add a monotonic constraint to this variable as we did for annual income in *Figure 8.18*:

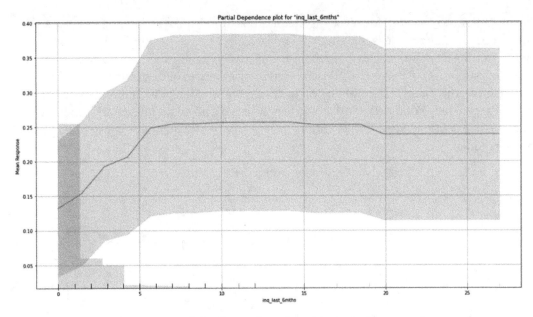

Figure 8.17 – PDP for the number of inquiries in the last 6 months

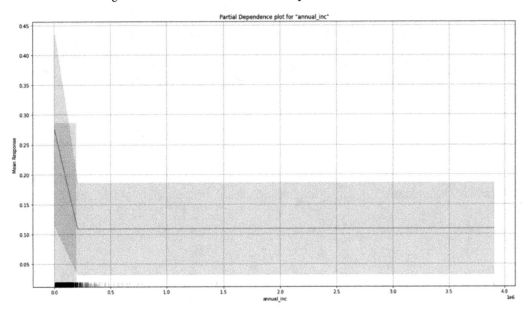

Figure 8.18 – PDP for monotonic annual income

Figure 8.19 shows the PDP for revolving credit utilization. Unlike earlier numeric plots, the `revol_util` variable is not visibly monotonic. In general, the higher the utilization, the greater the default rate. However, there is a relatively high default rate at the utilization of zero. Sometimes, effects such as this are caused by mixtures of disparate populations. For example, this could be a combination of customers who have credit lines but carry no balances (generally good risks) with customers who have no credit lines at all (generally poorer risks). Without reparameterization, `revol_util` should not be constrained to be monotonic:

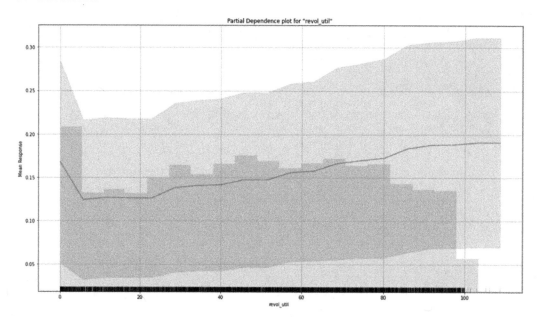

Figure 8.19 – PDP for revolving utilization

Finally, *Figure 8.20* shows the SHAP summary for the final model. The relative importance in terms of SHAP values is slightly different than that of our feature importance and PDP views:

SHAP Summary

SHAP summary plot shows the contribution of the features for each instance (row of data). The sum of the feature contributions and the bias term is equal to the raw prediction of the model, i.e., prediction before applying inverse link function.

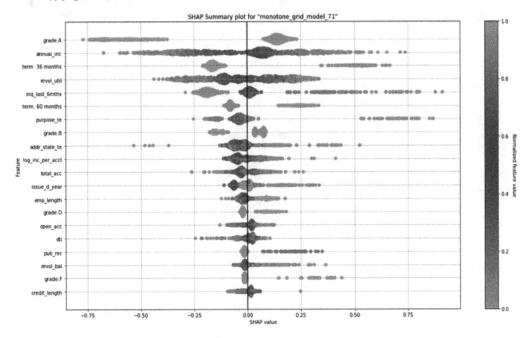

Figure 8.20 – The SHAP Summary plot for the final model

This has been a taster of what a final model review or whitepaper would show. Some of these are multiple pages in length.

Preparation for model pipeline deployment

Exporting a model as a MOJO for final model deployment is trivial, for instance, consider the following:

```
final.download_MOJO("final_MOJO.zip")
```

Deployment of the MOJO in various architectures via multiple recipes is covered, in detail, in *Chapter 9, Production Scoring and the H2O MOJO*. In general, there is a significant amount of effort that must be assigned to productionizing data for model scoring. The key is that data used in production must have a schema identical to that of the training data used in modeling. In our case, that means all the data wrangling and feature engineering tasks must be productionized before scoring in production can occur. In other words, the process is simply as follows:

1. Transform raw data into the training data format.

2. Score the model using the MOJO on the transformed data.

It is a best practice to work with your DevOps or equivalent production team well in advance of model delivery to understand the data requirements for deployment. This includes specifying roles and responsibilities such as who is responsible for producing the data transformation code, how is the code to be tested, who is responsible for implementation, and more. Usually, the delivery of a MOJO is not the end of the effort for a data science leader. We will discuss the importance of this partnership, in more detail, in *Chapter 9, Production Scoring and the H2O MOJO*.

Summary

In this chapter, we reviewed the entire data science model-building process. We started with raw data and a somewhat vaguely defined use case. Further inspection of the data allowed us to refine the problem statement to one that was relevant to the business and that could be addressed with the data at hand. We performed extensive feature engineering in the hopes that some features might be important predictors in our model. We introduced an efficient and powerful method of model building using H2O AutoML to build an array of different models using multiple algorithms. Selecting one of those models, we demonstrated how to further refine the model with additional hyperparameter tuning using grid search. Throughout the model-building process, we used the diagnostics and model explanations introduced in *Chapter 7, Understanding ML Models*, to evaluate our ML model. After arriving at a suitable model, we showed the simple steps required to prepare for the enterprise deployment of a model pipeline built in H2O.

The next chapter introduces us to the process of deploying these models into production using the H2O MOJO for scoring.

Section 3 – Deploying Your Models to Production Environments

In this section, we learn how to take models exported from model building and deploy them for scoring on enterprise systems. We first focus on the enterprise-grade qualities and technical nature of the H2O MOJO model scoring artifact. We get hands-on by writing a batch file scoring program that embeds the MOJO in the program. We then show how MOJOs are deployed to diverse enterprise systems by detailing over a dozen deployment patterns for real-time, streaming, and batch scoring on H2O software, third-party integrations, and custom-built systems.

This section comprises the following chapters:

- *Chapter 9, Production Scoring and the H2O MOJO*
- *Chapter 10, H2O Model Deployment Patterns*

9
Production Scoring and the H2O MOJO

We spent the entire previous section learning how to build world-class models against data at scale with H2O. In this chapter, we will learn how to deploy these models and make predictions from them. First, we will cover the background on putting models into production scoring systems. We will then learn how H2O makes this easy and flexible. At the center of this story is the H2O **MOJO** (short for **Model Object, Optimized**), a ready-to-deploy scoring artifact that you export from your model building environment. We will learn technically what a MOJO is and how to deploy it. We will then code a simple batch file scoring program and embed a MOJO in it. We will finish with some final notes on the MOJO. Altogether, in this chapter, you will develop the knowledge to deploy H2O models in diverse ways and so begin achieving value from live predictions.

These are the main topics we will cover in this chapter:

- Relating the model building context to the scoring context for H2O models
- Recognizing the diversity of target production systems for H2O models
- Examining the technical design of the H2O deployable artifact, the H2O MOJO
- Writing your own H2O MOJO batch file scorer to show how to embed MOJOs in your own software

Technical requirements

In this chapter, you will need a Java SE 8 or greater environment. A Java IDE such as Eclipse is optional but useful. You will get a MOJO, a dataset to score and the Java code for the batch file scorer program in the following GitHub repository: `https://github.com/PacktPublishing/Machine-Learning-at-Scale-with-H2O/tree/main/chapt9`. These artifacts were generated from the model built in *Chapter 8*, *Putting It All Together*.

Note that we are done with model building at this point, so you do not need a model building environment pointing to a running H2O cluster.

The model building and model scoring contexts

In *Section 2*, *Building State-of-the-Art Models on Large Data Volumes Using H2O*, we spent a great amount of focus on building world-class models at scale with H2O. Building highly accurate and trusted models against massive datasets can potentially generate millions of dollars for a business, save lives, and define new product areas, but only when the models are deployed to production systems where predictions are made and acted upon.

This last step of deploying and predicting (or scoring) on a production system can often be time-consuming, problematic, and risky for reasons discussed shortly. H2O makes this transition from a built (trained) model to a deployed model easy. It also provides a wide range of flexibility in regard to where scoring is done (device, web application, database, microservice endpoint, or Kafka queue) and to the velocity of data (real-time, batch, and streaming). And, whatever the production context, the H2O deployed model scores lightning fast.

At the center of this ease, flexibility, and low-latency production scoring is the H2O MOJO. An H2O MOJO is a ready-to-deploy scoring artifact that is generated by a simple export command at the end of your model-building code. H2O MOJOs are similar regardless of the model-building algorithm that generated them. As a result, all H2O models are deployed similarly. Before diving into the MOJO and learning how to deploy it, let's first take a look in general at the process of moving from model training to model scoring.

Model training to production model scoring

We'll first take a general view of how models transition from model training to production scoring and then see how this is done with H2O.

Generic training-to-scoring pipeline

A generic pipeline of a trained to a deployed model can be represented as follows:

Figure 9.1 – Generalized pipeline from model training to scoring

Do note that this pipeline is more formally represented and elaborated by the practice called **Machine Learning Operations** (**MLOps**), which involves a larger area of concern, but for the focus of deploying a model to production, the representation here should work for us.

Each step is summarized as follows:

1. **Trained model**: The trained model is converted into a format or software object that can be deployed to a software system. This conversion can occur in many ways. The riskiest and most time-consuming way is to determine the mathematical logic of the trained model (for example, the branching logic of a tree-based model) and then rewrite that into `if-else` logic in software code. This is time-consuming because the logic of the trained model must be accurately communicated by the data scientist to the software developer, who must implement the logic correctly and then have it tested thoroughly to validate its accuracy. This is also error-prone and, therefore, risky, as well as time-consuming.

   The best-case scenario is when a conversion tool translates the trained model into a deployable artifact. This can be either a format (for example, XML for PMML, PFA, or ONNX) that declares the logic for a production system that is ready to compute against the declarative constructs, or it can be a runnable software artifact (for example, a Python wheel or Java JAR file) that can be embedded into a software program or framework.

2. **Deployable model**: The converted model is deployed to a production system. This written code or converted artifact is integrated into a software application or framework that, at some point, inputs data into the scoring logic it holds and outputs the scoring result. For example, a customer's data goes in and the probability of churn comes out.

   Model deployment should be performed in TEST and **production** (**PROD**) environments, with deployment and promotion done through a formal governance process using a **continuous integration and continuous deployment** (**CI/CD**) pipeline, as with the deployment of software in general. Deployable artifacts that are recognizable and standardized across all models built (for example, among different ML algorithms) are easier to automate during deployment than those that are not.

3. **Production system**: Scoring live in production. Production scoring needs can be diverse. Scoring may be needed, for example, against entire database tables in one batch, against each live ATM transaction sent over the network, inside a web application for every web page click of a customer, or on streams of sensor data sent from edge devices. Scoring may be on a device or on a large server in the cloud. Typically, the faster the score, the better (demands of less than 50 microseconds per score or faster are not uncommon), and the smaller the scorer size and resource consumption footprint, the closer to the edge it can be deployed.

4. **Predictions**: Scoring the output. Models output predictions during scoring. Note that predictions need a business context and action to achieve purpose or value. For example, customers who are predicted to churn are given phone calls or special offers to help ensure they remain customers. Often, scoring outputs require not just predictions, but also explanations in the form of reason codes for those predictions. How did the model weigh each input to the scorer when generating the prediction for a particular customer? In other words, which factors were most important in a specific prediction. These decision weights are represented as reason codes and they can help personalize a phone call or special offer in the churn case.

Let's see how the training-to-scoring pipeline is realized with H2O.

The H2O pipeline and its advantages

Trained H2O models participate in a similar pipeline as discussed, but with important attributes that make them easy to deploy to a diverse target of software systems and are also very fast when they score there. The deployable artifact for H2O is called a MOJO and it bridges the gap between model training and model scoring, and so is the central character in the story. Attributes of the H2O pipeline are summarized as follows:

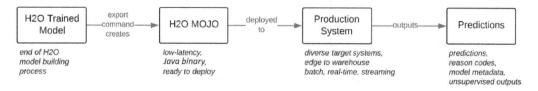

Figure 9.2 – H2O's model training-to-scoring pipeline

Let's elaborate on H2O's advantages of deploying models:

1. **H2O trained model**: H2O MOJOs exported from the model-building IDE as ready to deploy. The data scientist converts the trained model into an exported and ready-to-deploy MOJO by writing a single line of code in the IDE.

2. **H2O MOJO**: H2O MOJOs are standardized low-latency scoring artifacts and ready to deploy. The MOJO construct is standardized and shared by all model types and has its own runtime that embeds in any Java runtime. This means that all MOJOS (models) are identically embedded in any **Java virtual machine (JVM)** independent of the larger software and hardware context. MOJOs are also lightweight and can be deployed to nearly all infrastructure (except the smallest of edge devices). MOJOs are super fast at scoring and can handle any data velocity (real-time scoring, batch scoring, and streaming scoring).

3. **Production system**: H2O MOJOs flexibly deploy to a diversity of production systems. MOJOs deploy to a wide range of production systems. An overview of these systems and details of how MOJOS are deployed to them are given a bit later in this chapter.

4. **Predictions**: MOJOs can output a lot of information in their scoring. Inputs to MOJO return predictions in the form of class probabilities for classification, predicted numeric values for regression, and model-specific outcomes for unsupervised problems. Additionally, and optionally, MOJOs may return reason codes in the form of Shapley or K-LIME values, or other attributes such as leaf node assignments for a prediction.

Let's focus more on H2O production scoring specifically in the next section.

H2O production scoring

Models achieve their business value when they are put into production to make predictions (or generate unsupervised results for an unsupervised class of problems). We discuss, in this section, a more detailed view of the H2O pipeline from model building to production scoring.

End-to-end production scoring pipeline with H2O

Take a look at the following diagram showing an end-to-end H2O pipeline from model training to model deployment and production scoring:

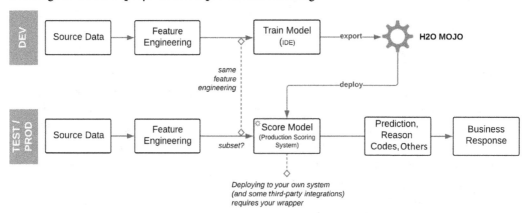

Figure 9.3 – High-level view of full scoring pipeline with H2O

Typically, model building is considered a **development** (**DEV**) environment, and model scoring is a PROD environment with source data from each respective environment.

For DEV, we have treated feature engineering and model training (and many associated steps such as model explainability and evaluation) extensively in *Section 2, Building State-of-the-Art Models on Large Data Volumes Using H2O*. We also briefly discussed the exportable ready-to-deploy H2O MOJO scoring artifact and deploying it to PROD systems earlier in this chapter.

Let's identify some key points to keep in mind during this pipeline:

- *You need feature engineering parity between DEV and PROD*: This means that any feature engineering done to create the training dataset must be matched by the scoring input in TEST/PROD. In other words, the features in the training dataset must be the same as those fed into the model scoring. If there were multiple steps of feature engineering (for example, **extract, transform, and load** (**ETL**) from a data source and feature engineering in H2O Sparkling Water) before constructing the training dataset in DEV, the input to scoring in TEST/PROD must have those same engineered features.

Having said that, depending on how the MOJO is deployed (H2O Scorer, third-party integration, or your own scorer), you likely will have to input to TEST/PROD only a subset of the features from those in the training dataset. This reflects the fact that the trained model typically selects only a subset of data features that contribute to the final model. This subsetting is not required, however; MOJOs can accept full or subsets of features (compared to the training dataset) depending on how you design it. This flexibility will become clearer later in the chapter when we take a closer look at deploying MOJOs.

- *You may need a wrapper around your MOJO (but not with H2O Scorers and most third-party integrations)*: MOJOs are ready to deploy to a Java environment. This means the MOJO is ready to convert input data to a prediction output using the mathematical logic derived from model training and held in the MOJO, and that the MOJO itself does not need compiling or modification in any way. But, you must still make sure the input (for example, CSV, JSON, batch, and so on) feeds into the MOJO in a way that the MOJO can accept. On the other side, you may want to extract more from the MOJO scoring result than only predictions, and you will need to convert the MOJO output to a format expected downstream in the application. You do this by writing a simple Java wrapper class and using the MOJO API called h2o-genmodel API to interact with the MOJO. These wrapper classes are not complicated. We will learn more about wrapping MOJOs with an example later in this chapter.

> **Important Note**
>
> H2O Scorers and many third-party integrations for MOJOs do not require wrappers because they handle this internally. All you need is the exported MOJO in these cases. Additionally, many integrations occur by way of REST APIs to endpoints of MOJOs deployed on REST servers.

- *You may want to return reason codes or other information with your predictions*: MOJOs return predictions for supervised models and model-specific output for unsupervised models (for example, an anomaly score is returned for a model trained with H2OIsolationForestEstimator. But, there is more to retrieve from the MOJO; you can also return reason codes as K-LIME or Shapley values, the decision path taken through tree-based models, or class labels for the prediction of classification problems. These additional outputs are implemented in wrapper code using the h2o-genmodel API for scorers you build. They may or may not be built into the functionality of H2O Scorers or out-of-the-box third-party integrations. You will need to check the specifications for these scorers.

- *You need a formalized process to deploy and govern your model*: Putting models into production involves risks: generally, the risk of failure or delay from errors during deployment, and the risk to revenue or reputation from adverse consequences from model decisions by deployed models. We will look at this topic more closely in *Chapter 14, H2O at Scale in a Larger Platform Context*.

- *You need MLOps to monitor your model*: Models in PROD typically need to be monitored to see whether values of input data are changing over time compared to those in the training data (this result is called data drift). In this case, the model may need to be retrained since the signal it was trained against has changed, possibly causing the predictive accuracy of the model to degrade. Bias, prediction distributions, and other aspects of scoring may also be monitored.

Model monitoring is outside the capability of MOJOs. MOJOs are concerned with single scores. Monitoring fundamentally tracks aggregate trends from MOJO inputs and outputs and is a separate area of technology and concern that will not be treated here. Do note, however, that H2O has an MLOps platform that performs model monitoring and governance. It is overviewed in *Chapter 16, The Machine Learning Life Cycle, AI Apps, and H2O AI Hybrid Cloud*.

We have just overviewed the full pipeline from H2O model building to production scoring and identified key points regarding this pipeline. One part of this pipeline is quite variable depending on your needs: the target system on which to deploy your MOJO. Let's explore this in greater detail.

Target production systems for H2O MOJOs

One large advantage of MOJOs is that they can be deployed to a wide range of production systems. Let's dig deeper using the following diagram to summarize:

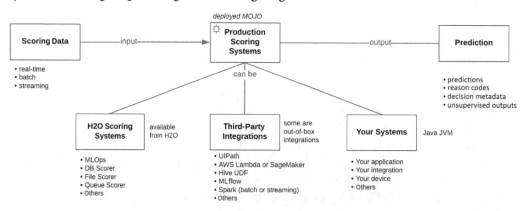

Figure 9.4 – Taxonomy of production systems for MOJO scoring

Business requirements mostly determine whether scoring needs to be real time, batch, or streaming and MOJOs can handle the full range of these data velocities.

It is useful to articulate production target systems into the following three categories for MOJO deployments:

- **H2O scoring system**: This represents the H2O scoring software that is available from H2O. These scorers include a REST server with MLOps and rich model monitoring and governance capabilities (and a lively roadmap that includes batch scoring, champion/challenger testing, A/B testing, and more), a database scorer for batch database table scoring that outputs to a table or file, a file batch scorer, and AMQ and Kafka Scorers for streaming events. H2O is actively adding more scorers, so visit their website to keep up to date. The MLOps scorer specifically is discussed in more detail in *Chapter 16, The Machine Learning Lifecycle, AI Apps, and H2O AI Hybrid Cloud.*

- **Third-party integrations**: Many third parties integrate out-of-the-box with MOJOs for scoring on their framework or software. Others require some glue to be built to create a custom integration.

- **Your own DIY system**: You can embed MOJOs in your software or framework integrations that run a Java environment. Integrations will require a simple Java wrapper class to interface your application or framework to the MOJO data input and output capabilities (for example, your REST server will need to convert JSON to a MOJO data object). H2O makes this easy with its **MOJO API**. Wrapping with the MOJO API is discussed in greater detail with code examples later in the chapter.

Note that this chapter provides an introduction to deploying MOJOs to target systems. The entire *Chapter 10, H2O Model Deployment Patterns,* will be devoted to walking through multiple examples of MOJO deployments to target systems.

Now that we understand the end-to-end H2O pipeline from model building to live scoring on diverse production systems, let's take a closer look at its central player: the MOJO.

H2O MOJO deep dive

All MOJOs are fundamentally similar from a deployment and scoring standpoint. This is true regardless of the MOJO's origin from an upstream model-building standpoint, that is, regardless of which of H2O's wide diversity of model-building algorithms (for example, Generalized Linear Model, and XGBoost) and techniques (for example, Stacked Ensembles and AutoML) and training dataset sizes (from GBs to TBs) were used to build the final model.

Let's get to know the MOJO in greater detail.

What is a MOJO?

A **MOJO** stands for Model Object, Optimized. It is exported from your model-building IDE by running the following line of code:

```
model.download_mojo(path="path/for/my/mojo")
```

This downloads a uniquely-named .zip file onto the filesystem of your IDE, to the path you specified. This .zip file is the MOJO and this is what is deployed. You do not unzip it, but if you are curious, it contains a model.ini file that describes the MOJO as well as multiple .bin files, all of which are used by the **MOJO runtime**.

What is a MOJO runtime? This is a Java .jar file called h2o-genmodel.jar and is a generic runtime for all H2O Core MOJOs. In other words, MOJOs are specific to the trained models they are derived from, and all MOJOs are loaded identically into the MOJO runtime. The MOJO runtime integrates with a Java runtime (in H2O software, third-party software, or your own software). The following diagram relates MOJOs to the MOJO runtime.

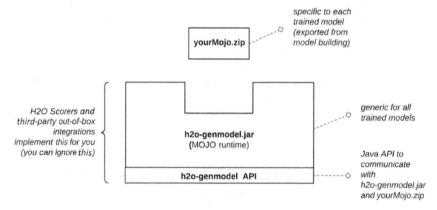

Figure 9.5 – MOJOs and the MOJO runtime

As mentioned previously, MOJOs are deployed to a Java runtime, more formally known as a **Java virtual machine** (**JVM**). The software that embeds the MOJO uses h2o-genmodel.jar as a dependent library to do so. The software loads the model-specific MOJO into the generic h2o-genmodel.jar runtime using the h2o-genmodel API. The actual scoring logic in the application code also uses h2o-genmodel.jar and its API to implement the scoring and extraction of results from the embedded MOJO.

Let's dig down and elaborate in the next section.

Deploying a MOJO

You need only the MOJO if you deploy a MOJO to an H2O Scorer or to a third-party software that integrates MOJOs out-of-the-box. You do not need to consider the MOJO runtime and API in these cases. This is because these software systems have already implemented h2o-genmodel.jar (using the h2o-genmodel API) behind the scenes, in other words, in the H2O Scorer or third-party software that is deployed and operating.

In other cases, you need to write the code that embeds the MOJO and extracts its scoring results. This code, typically, is a single Java wrapper class that uses the h2o-genmodel API. We will visit this a bit later using a code example.

This distinction is important and deserves a larger callout.

> **Key Distinction in MOJO Deployment**
>
> You need only the MOJO when deploying to H2O scoring software or third-party software that integrates MOJOs out of the box (configuration-based).
>
> You need to write a simple Java wrapper class using the h2o-genmodel API when integrating the MOJO into your own software or third-party software that does not integrate MOJO out of the box. This wrapper requires h2o-genmodel.jar, which is the library that the h2o-genmodel API represents.
>
> (If you are consuming MOJO predictions in third-party software or your own software from a REST server, you do not, of course, need the MOJO or the MOJO runtime. You simply need to conform to the REST endpoint API for the MOJO.)

Let's look at the case when you need to write a wrapper.

Wrapping MOJOs using the H2O MOJO API

Let's first touch upon a few precursors before learning how to wrap MOJOs inside larger software programs.

Obtaining the MOJO runtime

You can download h2o-genmodel.jar when you download your MOJO from the IDE after model building. This is simply a matter of adding a new argument to your download statement, as follows:

```
Model.download_mojo(path="path/for/my/mojo",
                    get_genmodel_jar=True)
```

This method of obtaining `h2o-genmodel.jar` generally is not done in a governed production deployment. This is because `h2o-genmodel.jar` is generic to all MOJOs and is a concern of the software developer and not the data scientists.

Software developers can download the MOJO runtime from the Maven repository at `https://mvnrepository.com/artifact/ai.h2o/h2o-genmodel`. The `h2o-genmodel.jar` is backward-compatible; it should work for a MOJO generated from an H2O-3 (or Sparkling Water) version equal to or less than the `h2o-genmodel.jar` version.

> **A Tip for Obtaining the MOJO Runtime (h2o-genmodel.jar)**
>
> Data scientists do not have to download the MOJO runtime each time they download their MOJO from their model-building IDEs. This is because the MOJO runtime is generic to all MOJOs. A best practice is to let your developers (not the data scientists) concern themselves with obtaining and using the MOJO runtime for production deployments when needed. This can be done through the Maven repository referenced earlier.

The h2o-genmodel API

Javadocs for the `h2o-genmodel` API are located at `https://docs.h2o.ai/h2o/latest-stable/h2o-genmodel/javadoc/index.html`. Note that this is for the latest H2O-3 (or Sparkling Water). To get a different version, go to `https://docs.h2o.ai/prior_h2o/index.html`.

In summary, the `h2o-genmodel` API is used to build wrappers around the MOJO so your application can feed data into the MOJO, extract prediction and decision information from it, and convert these results to the code in your wrapper. The wrapper is typically part of your larger application and can be seen as the glue between your app and the MOJO.

Let's dive in.

A generalized approach to wrapping your MOJO

It will be useful before writing code to first look at the logical flow of application code for the MOJO wrapper you develop. This can be seen in the following diagram:

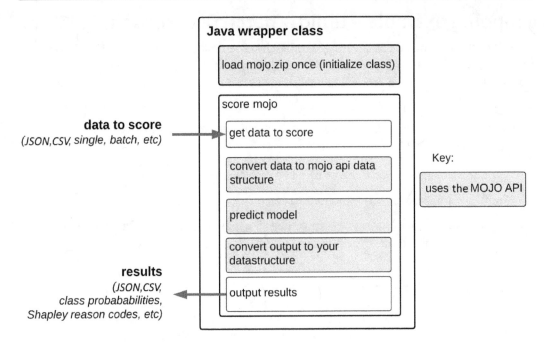

Figure 9.6 – Logical view of wrapping a MOJO

The Java wrapper typically is its own class (or part of a class) and imports
h2o-genmodel.jar and follows these general logical steps:

1. Load yourMOJO.zip into the MOJO runtime. Recall that h2o-genmodel.jar
    is the runtime that holds the generic logic to work on model-specific MOJOs. This
    runtime is now ready to operate on your specific model.

2. Feed data into the MOJO. To do so, convert the Java data structure of your input
    into a MOJO data structure using the h2o-genmodel code.

3. Score the MOJO. This is a single line of h2o-genmodel code.

4. Extract the subset of information that you need from the MOJO scoring results.
    Recall that prediction results (or unsupervised results) represent aspects of the
    prediction (labels and predictions) as well as aspects of scoring decisions (reason
    codes, decision path to leaf node results, and other).

5. Convert the extracted results into a data structure needed downstream by
    the application.

Let's write a wrapper.

Wrapping example – Build a batch file scorer in Java

The goal of the wrapper we are writing is to batch score new data from a file. The output of the scoring will be the input record, the prediction, and the reason codes all formatted as a line of CSV. The reason codes will be a single CSV field but the reason codes will be pipe-delimited.

We will compile this wrapper class as a runnable program that accepts three input parameters:

- Input param 1: `path/of/batch/file/to/score`

- Input param 2: `path/to/yourMOJO.zip`

- Input param 3 (optional): The `–shap` flag to trigger the return of Shapley reason codes in addition to the scoring prediction for each row in the file

> **Shapley Values Add Latency**
>
> Keep in mind that returning Shapley values adds additional computation and, therefore, latency to each scoring. You might want to benchmark latencies with and without Shapley reason codes in your results to evaluate whether to include them in scoring or not if latency is critical.

We will use the MOJO that you exported at the end of your model building exercise in *Chapter 8, Putting It All Together*.

The code

Our batch file scorer program will involve a single Java class and will not include error handling and other production quality software design. Our purpose here is to show the fundamentals of integrating a MOJO into your software.

Note that the code samples below are elaborated step by step. To access the entire Java code from beginning to end, go to the GitHub repository at `https://github.com/PacktPublishing/Machine-Learning-at-Scale-with-H2O/tree/main/chapt9`.

Let's get started:

1. **Create an empty wrapper class**: First, create a class called `BatchFileScorer`. Since this is also an executable program, we will create a `main` method to start the code execution. Note the `import` statements for the `h2o-genmodel` library packages:

```
Import java.io.*;
import hex.genmodel.easy.RowData;
import hex.genmodel.easy.EasyPredictModelWrapper;
import hex.genmodel.easy.prediction.*;
import hex.genmodel.MojoModel;
public class BatchFileScorer {
   public static void main(String[] args) throws
Exception{

   // we will fill with steps 2 to 4 that follows
   }
}
```

Now, let's fill the `main` method with code, as shown in the following steps.

2. **Retrieve the input parameters**: We retrieve the input parameters from the program's arguments:

```
// get input parameters
File fileToScore = new File(args[0]);
String pathToMojo = args[1];
boolean doShapley = args.length == 3
   && args[2].equals("--shap");
```

3. **Load the MOJO and configure it to optionally return reason codes**: We load the MOJO into the MOJO runtime and configure it to return Shapley values:

```
// Load the mojo (only once) and configure
EasyPredictModelWrapper.Config config =
  new EasyPredictModelWrapper.Config();

config.setModel(MojoModel.load(pathToMojo);
if (doShapley) config.setEnableContributions(true);

EasyPredictModelWrapper model =
  new EasyPredictModelWrapper(config);
```

The MOJO is loaded only once here before all scoring later in the code.

> **Important Design Point – Load Your MOJO Once**
>
> Loading the MOJO can take a few seconds, but it only needs to be loaded into your program once.
>
> Load the MOJO once in your wrapper class (for example, when it initializes) before making all scoring requests. You do not want your sub-hundred or sub-ten millisecond scores each preceded by multiple seconds of loading.

Now to the magic: generating predictions.

4. **Score**: Next, we open the file and iterate and score each line using the MOJO API provided by the import statements shown in *step 1*:

```
// get each record from the file
BufferedReader br = new BufferedReader(new
  FileReader(fileToScore));

// we are skipping the first line (header line)
br.readLine();
String record = null;
while ((record = br.readLine()) != null) {
  // Convert input record to type required by mojo api
  RowData mojoRow = convertInput(record);
  // make the prediction
  BinomialModelPrediction p = model.
predictBinomial(mojoRow);
```

```
    // get results from p and format it to your needs
    // in this case, format is csv to write to file
    String outputString = formatOutput(record, p,
doShapley);

    // can write this to file
    // but printing to screen for ease of code explanation
    System.out.println(outputString);
    }
```

That is it! You have loaded the MOJO and configured the scoring, and scored each line of a file. To score, you have converted each record from its application representation (CSV string) to the h2o-genmodel representation (the DataRow object). You have written one line of code to score the record. And, you have retrieved the prediction and, optionally, Shapley reason codes from the scoring result. You then formatted this to a representation used by your application.

Drill-downs to the code

Let's drill down into methods from the previous code.

Method drilldown – Converting your application data object to an h2o-genmodel data object

Note that RowData mojoRow is where the program code is converted into the h2o-genmodel API data object. In the example here, it is done through the convertInput(record) method as shown:

```
private static RowData convertInput(String record) {
    String[] featureValues = record.split(",");

    RowData row = new RowData();
    row.put("purpose_te", featureValues[0]);
    row.put("addr_state_te", featureValues[1]);
    row.put("loan_amnt", featureValues[2]);
    row.put("term", featureValues[3]);
    row.put("installment", featureValues[4]);
    row.put("grade", featureValues[5]);
    // omitting features 6 to 24, see code in github repo
```

```
    row.put("emp_length_missing", featureValues[25]);


    return row;
}
```

We have simply split the input using a comma as a separator and assigned each value to the H2O `RowData` object, which essentially is a map of key-value pairs with the keys representing feature names (that is, column headings). There are alternatives to using `RowData`.

> **Design Decision – Choices for converting Your Data Object to the MOJO API Data Object**
>
> Using the `h2o-genmodel` API's `RowData` class, as we did here, is just one way to convert your application data object into an `h2o-genmodel` object to feed to the MOJO for scoring. Check the API for additional ways that may offer better code design for your implementation.

Method drilldown – The single line to score

Only a single line of code was needed to score the MOJO and retrieve results:

```
    BinomialModelPrediction p = model.predictBinomial(mojoRow);
```

Note that you may need a different class than `BinomialModelPrediction` depending on which type of model you build. Check the `h2o-genmodel` Javadocs for details on which Java class to use and what scoring information is returned.

Method drilldown – Collecting results and formatting as output

We ultimately constructed a string from the scoring results using the `formatOutput(record, p, doShapley)` method. Here is how that method was implemented:

```
private static String formatOutput(String record,
    BinomialModelPrediction p, boolean doShapley) {
    // start the ouput string with the record being scored
    String outputString = record;
    // add prediction to output string
    outputString += "   PREDICTION (good=0, bad=1): " + p.label
    + " " + p.classProbabilities[0];
```

```
    // add Shapley values (bar-delimited) to output string
  if(doShapley) {
    outputString += "  SHAP VALUES > 0.01: ";

    for (int i=0; i < p.contributions.length; i++) {
        // retrieving only Shap values over 0.01
        if (p.contributions[i] <  0.01) continue;
        outputString += model.getContributionNames()[i] + ": "
        + p.contributions[i] + "|" ;
    }

    return outputString;
  }
```

The main point here is that the prediction results are held in the h2o-genmodel API's BinomialModelPrediction p object that was returned from scoring. We can retrieve a lot of information from this object. In our case, we retrieved the predicted class, identified by p.label, and its probability, p.classProbabilities[0]. Since this is a BinomialModelPrediction, the probability of the other class would be retrieved by p.classProbabilities[1].

We then iterated through an array of the Shapley reason contribution names (model. getContributionNames()[i]) and values (p.contributions[i]). In our case, we are retrieving only reason codes with values over 0.01. Alternatively, for example, we could have sorted the reasons by value and returned the top five. When returning all reasons, a bias is returned as the last in the array, and the sum of all features and the bias will equal the raw prediction of the model.

Altogether, we used a bunch of code to format all of this into a CSV string starting with the original record and then appending the predicted class and its probability, and then a bar-delimited list of reason codes.

Running the code

To run the application, compile `BatchFileScorer.java` with `h2o-genmodel.jar` as an executable JAR file called `BatchFileScorer.jar`. Then, run the following command in the same directory as `BatchFileScorer.jar`:

```
java -jar BatchFileScorer.jar \
path/to/file/to/score \
path/to/mojo
```

To retrieve Shapley reason codes, append `--shap` to the statement.

Other things to know about MOJOs

You are now ready to deploy MOJOs, with or without required wrappers, as articulated in the previous section. Let's round up our knowledge of MOJOs by addressing the following secondary topics.

Inspecting MOJO decision logic

For tree-based models, you can use a utility built into `h2o-genmodel.jar` to generate a graphical representation of the tree logic in the MOJO. Here is how.

Let's use the same MOJO we used in the previous coding example of building a wrapper class. On the command line where your `h2o-genmodel.jar` is located, run the following:

```
java -cp h2o-genmodel.jar hex.genmodel.tools.PrintMojo \
-i "path/to/mojo" \
-o tree.png \
--format png \
--tree 0
```

This will create a .png file that looks like this:

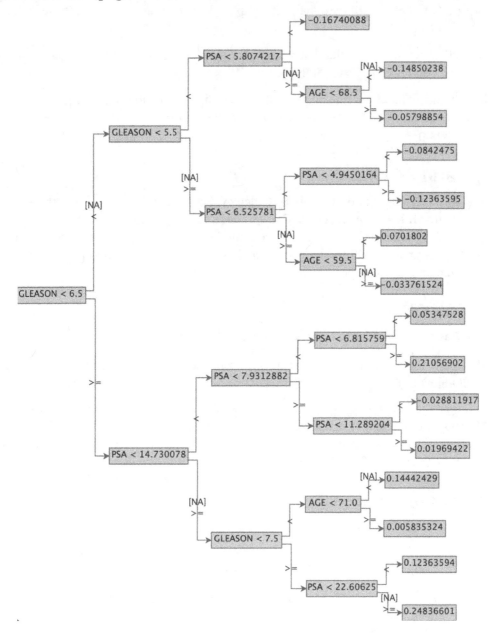

Figure 9.7 – Output of the PrintMojo utility

Note that if you omitted --tree 0, you would have generated a folder holding a forest of all trees. We have specified to return only the first one.

You can also use dot for --format. This produces a format that can be consumed by the third-party **Graphviz** utility to make the graphical representation more prettified than that shown in *Figure 9.7*.

Alternatively, if you wish to include this output for programmatic use, for -format, state .json, which outputs the file to JSON format.

See the H2O documentation for more details and configuration alternatives: https://docs.h2o.ai/h2o/latest-stable/h2o-docs/productionizing.html#viewing-a-mojo-model.

MOJO and POJO

OK, let's say it: MOJOs are not the only H2O deployable artifact. Before MOJOs, there were only **Plain Old Java Objects** (**POJOs**). MOJOs and POJOs are similar from a deployment viewpoint; they are generated from model building, are ready to deploy, and use the h2o-genmodel API to build wrapper classes, as we discussed before. They are also a bit different. Let's compare, contrast, and conclude.

MOJO and POJO similarities

The following are the similarities between MOJOs and POJOs:

- They are both exported from the IDE after your model is built (or from the H2O Flow UI).

- They are both deployed in the same way: they both run in a JVM, there is the wrapper or no wrapper distinction depending on the target scoring system (H2O Scorers, third-party, or your own software program), and they both use the MOJO runtime (h2o-genmodel.jar) and the same API and Javadoc.

MOJO and POJO differences

These are the differences between MOJOs and POJOs:

- A POJO is exported as a single .java file that needs to be compiled, whereas the MOJO exports as a single .zip file, as described earlier.

- POJOs contain entire trees to navigate the model, whereas MOJOs contain tree metadata and use generic tree-walker code in h2o-genmodel.zip to navigate the model. The larger the tree structure, the larger the POJO.

- POJOs are significantly larger than MOJOs (typically 20-25 times larger) and slower than MOJOs when scoring (2-3 times slower). In general, the larger the POJO, the slower it is compared to any MOJO built from the same model.

- Large POJOs may have trouble compiling. POJOs over 1 GB are not supported by H2O.

When to use either a MOJO or POJO

You should view POJOs as deprecated but still supported (except > 1GB) and sometimes needed in edge cases. Know that MOJOs are not fully supported across all algorithms so, in these cases, you are forced to use POJOs. Therefore, use MOJOs when you can and resort to POJOs in infrequent cases when you cannot.

> **Deployment Decision – MOJO or POJO?**
>
> See MOJOs as your go-to current technology and POJOs as similar to deploy but deprecated yet supported (except > 1 GB). MOJOs have advantages primarily in scoring speed and size footprint.
>
> MOJOs are not supported for some algorithms. Check the H2O documentation for current support considerations for MOJOs and POJOs.

We are now ready to summarize.

Summary

We began this chapter by taking a high-level view of the transition from model building to model deployment. We saw that this transition is bridged for H2O by the MOJO, a deployable representation of the trained model that is easy to generate from model building and easy to deploy for fast model scoring.

We then took a closer look at the range of target systems MOJOs can be deployed on, and saw that these must run in a Java runtime but, otherwise, are quite diverse. MOJOs can be scored on real-time, batch, and streaming systems, usefully categorized as H2O Scorers (scoring software provided and supported by H2O), third-party integrations (software provided and supported by companies other than H2O), and your software integrations (software that you build and maintain).

This categorization of target systems helps us determine whether you can deploy the exported MOJO directly, or whether you need to wrap it in a Java class using the h2o-genmodel API to embed it into the scoring software. H2O Scorers and some third-party scorers require only the exported MOJO and no wrapper to be implemented.

We then took a detailed look at the MOJO and the MOJO runtime, and how these relate to deployments with and without the need for wrappers. We described the general structure of a MOJO wrapper and coded a wrapper to batch score records from a file. Our coding gave us a better understanding of the MOJO API that is used to interact with the MOJO in your application. This understanding included how to use the API to load the MOJO, structure data to a type that can be used by the MOJO, score with the MOJO, and retrieve predictions and reason codes from the scoring results.

We then learned how to use a handy tool in the MOJO API to obtain a visual, JSON, or dot representation of the decision logic in the MOJO for your model.

Finally, we introduced the predecessor of the MOJO, the POJO, and characterized it as similar to the MOJO in terms of deployment and use of the MOJO API but deprecated yet supported, and so to be used for a minority of cases when MOJOs cannot.

Now, we understand in great detail the MOJO and how it is flexibly deployed to a diversity of production scoring systems. Let's move to the next chapter where we will exhibit this flexibility and diversity by describing concrete MOJO deployments on a handful of these systems.

10
H2O Model Deployment Patterns

In the previous chapter, we learned how easy it is to generate a ready-to-deploy scoring artifact from our model-building step and how this artifact, called a MOJO, is designed to flexibly deploy to a wide diversity of production systems.

In this chapter, we explore this flexibility of MOJO deployment by surveying a wide range of MOJO deployment patterns and digging down into the details of each deployment pattern. We will see how MOJOs are implemented for scoring on either H2O software, third-party software including **business intelligence** (**BI**) tools, and your own software. These implementations will include scoring on real-time, batch, and streaming data.

Recall from *Chapter 1*, *Opportunities and Challenges*, how **machine learning** (ML) models achieve business value when deployed to production systems. The knowledge you gain in this chapter will allow you to find the appropriate MOJO deployment pattern for a particular business case. For example, it will allow analysts to perform time-series forecasting from a **Microsoft Excel** spreadsheet, technicians to respond to predictions of product defects made on data streaming from a manufacturing process, or business stakeholders to respond to fraud predictions scored directly on Snowflake tables.

The goal of this chapter is for you to implement your own H2O model scoring, whether from these examples, your web search, or your imagination, inspired by these examples.

So, in this chapter, we're going to cover the following main topics:

- Surveying a sample of MOJO deployment patterns
- Exploring examples of MOJO scoring on H2O software
- Exploring examples of MOJO scoring on third-party software
- Exploring examples of MOJO scoring on your target-system software
- Exploring examples of accelerators based on H2O Driverless AI integrations

Technical requirements

There are no technical requirements for this chapter, though we will be highlighting the technical steps to implement and execute MOJO deployment patterns.

Surveying a sample of MOJO deployment patterns

The purpose of this chapter is to overview the diverse ways in which MOJOs can be deployed for making predictions. Enough detail is given to provide an understanding of the context of MOJO deployment and scoring. Links are provided to find low-level details.

First, let's summarize our sample of MOJO scoring patterns in table form to get a sense of the many different ways you can deploy MOJOs. After this sample overview, we will elaborate on each table entry more fully.

Note that the table columns for our deployment-pattern summaries are represented as follows:

- **Data Velocity**: This refers to the size and speed of data that is scored and is categorized as either **real-time** (single record scored, typically in less than 100 milliseconds), **batch** (large numbers of records scored at one time), and **streaming** (a continuous flow of records that are scored).
- **Scoring Communication**: This refers to how the scoring is triggered and communicated—for example, via a **REpresentational State Transfer** (**REST**) call or a **Structured Query Language** (**SQL**) statement.
- **MOJO Deployment**: This is a brief description of how the MOJO is deployed on the scoring system.

Let's take a look at some of the deployment patterns. We will break these patterns into four categories.

H2O software

This is a sample of ways you can deploy and score MOJOs on software provided and supported by H2O.ai. The following table provides a summary of this:

Name	Velocity	Communication	MOJO Deployment
H2O MLOps	Real-time; batch; streaming	REST	Upload MOJO
H2O eScorer	Real-time; batch	REST	Upload MOJO
H2O batch database scorer	Batch	SQL	Upload MOJO
H2O batch file scorer	Batch	File listener	Upload MOJO
H2O Kafka scorer	Streaming	Kafka topic	Upload MOJO
H2O batch scoring on Spark	Batch	`spark-submit`	Copied during `spark-submit`

We will see that deploying to H2O software is super easy since all you have to do is upload the MOJO (manually or programmatically).

Third-party software integrations

Here are a few examples of MOJO scoring with third-party software:

Name	Data Velocity	Scoring Communication	MOJO Deployment
Snowflake data warehouse	Batch; real-time	SQL; Snowpark **application programming interface (API)**	Snowflake integration
Teradata analytics	Batch; real-time	SQL	Teradata integration
MS Excel	Real-time	REST	REST server
Tableau	Real-time	REST	REST server
MS Power BI	Real-time	REST	REST server
UiPath **robotic process automation (RPA)**	Real-time	REST	REST server

Note that some third-party integrations are done by consuming scoring from MOJOs deployed to a REST server. This has the advantage of centralizing your deployment in one place (the REST server) and consuming it from many places (for example, dozens of Tableau or MS Excel instances deployed on employee personal computers).

Other third-party integrations are accomplished by deploying MOJOs directly on the third-party software system. The Snowflake integration, for example, is implemented on the Snowflake architecture and allows batch scoring that performs at a Snowflake scale (it can score hundreds of thousands of rows per second).

Your software integrations

We will explore the following patterns for integrating MOJOs directly into your own software:

Name	Data Velocity	Communication	MOJO Deployment
Software application	Real-time	Your design	MOJO with a wrapper class
On-device scoring	Real-time; streaming	On-device	MOJO with a wrapper class

MOJO integration into your software requires a MOJO wrapper class. We learned how to do this in *Chapter 9, Production Scoring and the H2O MOJO*. Of course, you can take the alternative approach and integrate your software with MOJO scoring consumed from a REST endpoint.

Accelerators based on H2O Driverless AI integrations

This book focuses on H2O Core (H2O-3 and Sparkling Water) model-building technology for building models against large data volumes. H2O provides an alternative model-building technology called Driverless AI. Driverless AI is a specialized, **automated ML (AutoML)** engine that allows users to find highly accurate and trusted models in extremely short amounts of time. Driverless AI cannot train on the massive datasets that H2O Core can, though. However, Driverless AI also produces a MOJO, and its flavor of MOJO deploys similarly to the H2O Core MOJO. These similarities were covered in *Chapter 9, Production Scoring and the H2O MOJO*.

There are many examples available online for deploying Driverless AI MOJOs. These examples can be followed as a guide to deploying H2O Core MOJOs in the same pattern. Consider the following Driverless AI examples therefore as accelerators that can get you most of the way to deploying your H2O Core MOJOs, but some implementation details will differ:

Name	Data Velocity	Communication	MOJO Deployment
Apache NiFi	Real-time; batch	NiFi processor	NiFi processor configuration
Apache Flink	Streaming; batch	Flink program	Flink program
Amazon Web Services (AWS) Lambda	Real-time; microbatch	REST	MOJO with a wrapper class
AWS SageMaker	Real-time	REST	MOJO with a wrapper class

The patterns shown in these four tables should provide a good sense of the many ways you can deploy MOJOs. They do not, however, represent the total set of possibilities.

> **A Note on Possibilities**
>
> The patterns shown here are merely a sample of H2O MOJO scoring patterns that exist or are possible. Other MOJO scoring patterns can be found through a web search, and you can use your imagination to integrate MOJO scoring in diverse ways into your own software. Additionally, H2O.ai is rapidly expanding its third-party partner integrations for scoring, as well as expanding its own MOJO deployment, monitoring, and management capabilities. This is a rapidly moving space.

Now that we have surveyed a landscape of MOJO deployment patterns, let's jump in and look at each example in detail.

Exploring examples of MOJO scoring with H2O software

The patterns in this section represent MOJOs deployed to H2O software. There are many advantages to deploying to H2O software. First, the software is supported by H2O and their team of ML experts. Second, this deployment workflow is greatly streamlined for H2O software since all you have to do is supply the MOJO in a simple upload (via a **user interface (UI)**, an API, or a transfer method such as remote copy). Third, H2O scoring software has additional capabilities—such as monitoring for prediction and data drift—that are important for models deployed to production systems.

Let's start by looking at H2O's flagship model-scoring platform.

H2O MLOps

H2O MLOps is a full-featured platform for deploying, monitoring, managing, and governing ML models. H2O MLOps is dedicated to deploying models at scale (many models and model versions, enterprise-grade throughput and performance, **high availability**, and so on), and addressing monitoring, management, and governance concerns around models in production.

H2O MLOps and its relation to H2O's larger **end-to-end** ML platform will be reviewed in *Chapter 13, Introducing H2O AI Cloud*. See also `https://docs.h2o.ai/mlops-release/latest-stable/docs/userguide/index.html` for the MLOps user guide to better understand H2O MLOps.

Pattern overview

The H2O MLOps scoring pattern is shown in the following diagram:

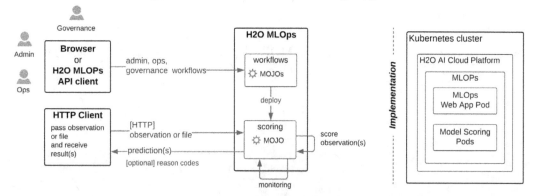

Figure 10.1 – Model-scoring pattern for H2O MLOps

We'll elaborate on this next.

Scoring context

This is H2O.ai's flagship model-deployment, model-monitoring, and model-governance platform. It can be used to host and score both H2O and third-party (non-H2O) models.

H2O MLOps scores models in real time and in batches. Predictions optionally return reason codes. Models are deployed as single-model, champion/challenger, and A/B. See the *Additional notes* section for a full description of its capabilities.

Implementation

H2O MLOps is a modern Kubernetes-based implementation deployed using Terraform scripts and Helm charts.

Scoring example

The following code snippet shows a real-time scoring request sent using the `curl` command:

```
curl -X POST -H "Content-Type: application/json" -d @- https://
model.prod.xyz.com/9c5c3042-1f9a-42b5-ac1a-9dca19414fbb/model/
score << EOF
{"fields":["loan_amnt","term","int_rate","emp_length","home_
ownership","annual_inc","purpose","addr_state","dti","delin-
q_2yrs","revol_util","total_acc","longest_credit_length","veri-
fication_status"rows":[["5000","36months","10.65","10",24000.0"
,"RENT","AZ","27.650","0","83.7","9","26","verified"]]}EOF
```

And here is the result:

```
{"fields":["bad_loan.0","bad_loan.1"],"id":"45d0677a-
9327-11ec-b656-2e37808d3384","sc
ore":[["0.7730158252427003","0.2269841747572997"]]}
```

From here, we see the probability of a loan default (`bad_loan` value of 1) is `0.2269841747572997`. The `id` field is used to identify the REST endpoint, which is useful when models are deployed in champion/challenger or A/B test modes.

Additional notes

Here is a brief summary of key H2O MLOps capabilities:

- **Multiple deployment models**: Standalone; champion/challenger; A/B models

- **Multiple model problems**: Tabular; time-series; image; language models

- **Shapley values**: On deployment, specify whether to return Shapley values (reason codes) with the prediction

- **Third-party models**: Scores and monitors non-H2O models—for example, scikit-learn models

- **Model management**: Model registry; versioning; model metadata; promotion and approval workflow

- **APIs**: APIs and **continuous integration and continuous delivery (CI/CD)** integration

- **Analytics**: Optionally push scoring data to your system for your own analytics

- **Lineage**: Understand the lineage of data, experiment, and model

- **Model monitoring**: Data drift and prediction monitoring with alert management (bias and other types of monitoring are on the MLOps roadmap)

> **H2O MLOps versus Other H2O Model-Scoring Software**
>
> MLOps is H2O's flagship full-featured platform to deploy, monitor, and govern models for scoring. H2O supplies other software (overviewed next) that is specialized to address needs or constraints where MLOps may not fit.

Next, let's have a look at the H2O REST scorer.

H2O eScorer

H2O has a lightweight but powerful REST server to score MOJOs, called the H2O eScorer. This is a good alternative for serving MOJOs as REST endpoints without committing to larger infrastructure requirements of the H2O MLOps platform and therefore freeing deployment options to on-premises and lightweight deployments. Recall that third-party software often integrates with MOJOs by way of REST endpoint integration, so this is an effective way to achieve that.

Pattern overview

The H2O REST scorer pattern is shown in the following diagram:

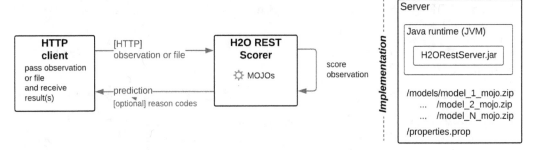

Figure 10.2 – MOJO scoring pattern for H2O REST scorer

Here is an elaboration.

Scoring context

The H2O REST scorer makes real-time and batch predictions to a REST endpoint. Predictions optionally include reason codes.

Implementation

The H2O Rest scorer is a single **Java ARchive (JAR)** file holding an Apache Tomcat server hosting a Spring REST services framework. A properties file configures the application to host multiple REST scoring endpoints. MOJOs are loaded either by REST itself or by other means of transferring the MOJO to the server.

High throughput is achieved by placing multiple H2O REST scorers behind a load balancer.

Scoring example

Here are some examples of REST endpoints for real-time scoring:

```
http://192.1.1.1:8080/model?name=riskmodel.mojo &row=50
00,36months,10.65,162.87,10,RENT,24000,VERIFIED-income,
AZ,27.65,0,1,0,13648,83.7,0"
```

The REST scorer's REST API is quite flexible. For example, it includes multiple ways to structure the payload (for example, an observation input can be sent as **comma-separated values (CSV)**, **JavaScript Object Notation (JSON)**, or other structures with the scorer output returned in the same format, which is convenient when integrating with a BI tool).

Additional notes

Here is a summary of the H2O REST Scorer's set of capabilities:

- Each H2O Rest scorer can score multiple models (that is, MOJOs), each with its own REST endpoint.

- Typically, 1,000 scores per second are achieved for each CPU on an H2O REST scorer server.

- Security, monitoring, and logging settings are configurable in a properties file.

- **Java Monitoring Beans (JMX)** can be configured so that your own monitoring tool can collect and analyze runtime statistics. Monitoring includes scoring errors, scoring latency, and data drift.

- Security features include **HTTPS**, administrator authentication, authenticated endpoint **URIs** and limited access from IP prefix.

- There is extensive logging.

- There are extensive capabilities via the REST API, including obtaining model metadata, defining prediction output formats, defining logging verbosity, and managing MOJOs on the server.

- The REST API can generate an example request sent from different BI tools to score a model on the H2O REST scorer—for example, sample Python code to call a model for Power BI.

Next, we will have a look at the H2O batch database scorer.

H2O batch database scorer

The H2O batch database scorer is a client application that can perform batch predictions against tables using a **Java Database Connectivity (JDBC)** connection.

Pattern overview

The H2O batch database scorer pattern is shown in the following diagram:

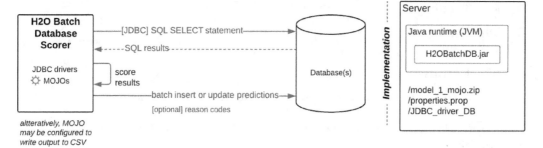

Figure 10.3 – MOJO scoring pattern for H2O batch database scorer

We'll elaborate on this next.

Scoring context

The H2O batch database scorer performs batch predictions against database tables. Predictions optionally include reason codes. Depending on how it is configured, predictions against table rows can be inserted into a new table or updated into the same table being scored. Alternatively, it can generate a CSV file of the prediction's outcome. This CSV output can be used to manually update tables or for other downstream processing.

Details of the processing sequence for H2O batch database scoring are shown in *Figure 10.3*.

Implementation

The H2O batch database scorer is a single JAR file that is available from H2O.ai. The JAR file uses a properties file to configure aspects of the database workflow.

More specifically, the property file contains the following:

- SQL connection string
- SQL SELECT statement to batch-score
- SQL INSERT or UPDATE statement to write prediction results
- Number of threads during batch scoring

- Path to MOJO
- Flag to write results to CSV or not
- Security settings
- Other settings

Scoring example

The following command shows how a batch job is run from the command line:

```
java -cp /PostgresData/postgresql-42.2.5.jar:H2OBatchDB.jar \
ai.h2o.H2OBatchDB
```

This, of course, can be integrated into a scheduler or a script to schedule and automate batch scores.

Note that this command does not include anything about the database or table. The program that is kicked off from this command finds the properties file, as described in the previous *Implementation* subsection, and uses the information there to drive batch scoring.

Additional notes

A single properties file holds all the information needed to run a single batch-scoring job (the properties file maps to a SQL statement against a table that will be scored).

If no properties file is stated in the Java command to score (see the *Scoring example* section), then the default properties file is used. Alternatively, a specific properties file can be specified in the Java command line to run a non-default scoring job.

Next, let's have a look at the H2O batch file scorer.

H2O batch file scorer

The H2O batch file scorer is an application that can perform batch predictions against records in a file.

Pattern overview

The H2O batch file scorer pattern is shown in the following diagram:

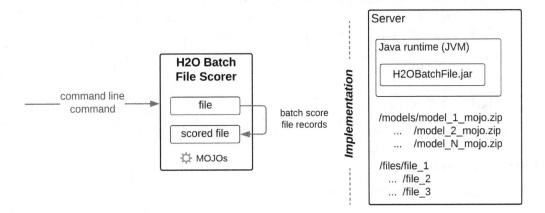

Figure 10.4 – MOJO scoring pattern for H2O batch file scorer

This is how it is used.

Scoring context

Scoring is batch against records in a file, and the output will be a file identical to the input file but with a scored field appended to each record. The output file remains on the H2O batch-scorer system until processed by another system (for example, copied to a downstream system for processing).

Implementation

The H2O batch file scorer is a single JAR file that is available from H2O.ai. Command-line arguments are used to specify the location of the model and input file, as well as any runtime parameters such as skipping the column head if one exists in the file.

Scoring example

The following command shows how a batch-file job is run from the command line:

```
java -Xms10g -Xmx10g -Dskipheader=true -Dautocolumns=true
-classpath mojo2-runtime.jar:DAIMojoRunner_TQ.jar
daimojorunner_tq.DAIMojoRunner_TQ pipeline.mojo LoanStats4.csv
```

A few notes are worth mentioning.

Additional notes

This scorer is ideal for processing extremely large files (> GB) as a single task, making it easy to use in a traditional batch-processing workflow. If the input file contains a header, then the scorer will select the correct columns to pass to the model, and if a header is not present, then the columns can be passed as command-line parameters.

Let's now take a look at the H2O Kafka scorer.

H2O Kafka scorer

The H2O Kafka scorer is an application that integrates with the score from Kafka streams.

Pattern overview

The H2O Kafka scorer pattern is shown in the following diagram:

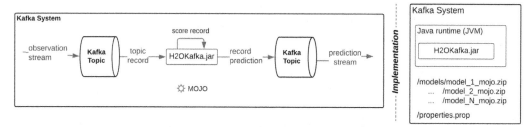

Figure 10.5 – MOJO scoring pattern for H2O Kafka scorer

Scoring context

Scoring against streaming data is shown in *Figure 10.5*. Specifically, the H2O Kafka scorer pulls messages from a topic queue and publishes the score outcome to another topic.

Implementation

The H2O Kafka scorer is a JAR file that is implemented on the Kafka system. A properties file is used to configure which topic to consume (and thus which messages to score) and which to publish to (where to send the results). When the H2O Kafka scorer JAR file is started, it loads the MOJO and then listens for incoming messages from the topic.

Scoring example

Scoring is done when a message arrives at the upstream topic. A prediction is appended to the last field of the original message. This new message is then sent to a topic for downstream processing.

Additional notes

Scaling throughput is done using native Kafka scaling techniques inherent in its distributed parallelized architecture.

Finally, let's look at H2O batch scoring on Spark.

H2O batch scoring on Spark

H2O MOJOs can be deployed as native Spark jobs.

Pattern overview

The H2O batch scoring on Spark pattern is shown in the following diagram:

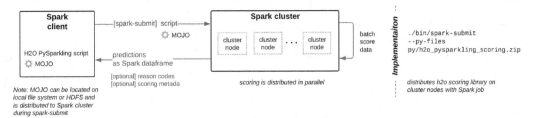

Figure 10.6 – MOJO scoring pattern for H2O batch scoring on Spark

Scoring context

Scoring is batch and on a Spark cluster. As such, the batch scoring is distributed and thus scales well to massive batch sizes.

Implementation

The required dependency to score MOJOs on the Spark cluster is distributed with the spark-submit command, as shown in the following section.

Scoring example

First, we'll create a **PySparkling** job similar to the following example. We will call this job myRiskScoring.py. The code is illustrated in the following snippet:

```
from pysparkling.ml import *
settings = H2OMOJOSettings(convertUnknownCategoricalLevelsToNa
= True, convertInvalidNumbersToNa = True)
model_location="hdfs:///models/risk/v2/riskmodel.zip"
model = H2OMOJOModel.createFromMojo(model_location, settings")
```

```
predictions = model.transform(dataset)
// do something with predictions, e.g. write to hdfs
```

Then, submit your Spark job with the H2O scoring library, as follows:

```
./bin/spark-submit \
    --py-files py/h2o_pysparkling_scoring.zip \
    myRiskScoring.py
```

Note that the h2o_pysparkling_scoring.zip dependency will be distributed to the cluster with the job. This library is available from H2O.ai.

Additional notes

There are other scoring settings available in addition to those shown in the previous code sample. The following link will provide more details: https://docs.h2o.ai/ sparkling-water/3.1/latest-stable/doc/deployment/load_mojo. html.

We have finished our review of some scoring patterns on H2O software. Let's now transition to scoring patterns on third-party software.

Exploring examples of MOJO scoring with third-party software

Let's now look at some examples of scoring that involve third-party software.

Snowflake integration

H2O.ai has partnered with Snowflake to integrate MOJO scoring against Snowflake tables. It is important to note that the MOJO in this integration is deployed on the Snowflake architecture and therefore achieves Snowflake's native scalability benefits. Combined with the low latency of MOJO scoring, the result is batch scoring on massive Snowflake tables in mere seconds, though real-time scoring on a smaller number of records is achievable as well.

Pattern overview

The Snowflake integration pattern is shown in the following diagram:

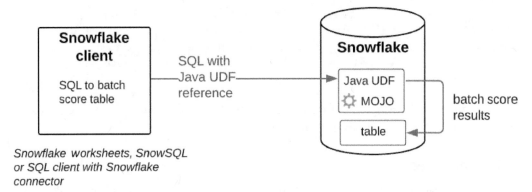

Figure 10.7 – MOJO scoring pattern for Snowflake Java user-defined function (UDF) integration

Let's elaborate.

Scoring context

Scoring is batch against Snowflake tables and leverages the scalability of the Snowflake platform. Thus, scoring can be made against any Snowflake table, including those holding massive datasets.

Scoring is done by running a SQL statement from a Snowflake client. This can be either a native Snowflake worksheet, SnowSQL, or a SQL client with a Snowflake connector. Alternatively, scoring can be done programmatically using Snowflake's Snowpark API.

Implementation

To implement your score, create a staging table and grant permissions against it. You then copy your MOJO and H2O JAR file dependencies to the staging table.

You can then use SQL to create a Java UDF that imports these dependencies and assigns a handler to the H2O dependency that does the scoring. This UDF is then referenced when making a SQL scoring statement, as shown next.

You can find H2O dependencies and instructions here: https://s3.amazonaws.com/artifacts.h2o.ai/releases/ai/h2o/dai-snowflake-integration/java-udf/download/index.html.

An integrated experience of using the UDF with Snowflake is also available online at https://cloud.h2o.ai/v1/latestapp/wave-snowflake.

Scoring example

This is an example of a SQL statement that performs batch scoring against a table:

```
select ID, H2OScore_Java('Modelname=riskmodel.zip', ARRAY_
CONSTRUCT(loan_amnt, term, int_rate, installment, emp_length,
annual_inc, verification_status, addr_state, dti, inq_
last_6mths, revol_bal, revol_util, total_acc)) as H2OPrediction
from RiskTable;
```

Notice that the H2O Scoring UDF (loaded as shown in the *Implementation* section) is run and that the model name (the MOJO name) is referenced.

Additional notes

For a more programmatic approach, you can use the Snowpark API instead of a SQL statement to batch-score.

Alternative implementation – Scoring via a Snowflake external function

For cases where you do not want to deploy MOJOs directly to the Snowflake environment, you can implement an external function on Snowflake and then pass the scoring to an H2O eScorer implementation. Note that scoring itself is external to Snowflake, and batch throughput rates are determined by the H2O eScorer and not the Snowflake architecture. This is shown in the following diagram:

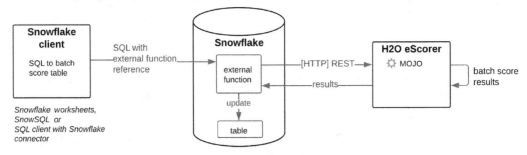

Figure 10.8 – MOJO scoring pattern for Snowflake external function integration

To implement this, we will use Snowflake on AWS as an example. Follow these steps:

1. First, use the Snowflake client to create `api_integration` to `aws_api_gateway`. A gateway is required to secure the external function when communicating to the H2O eScorer, which will be outside Snowflake. You will need to have the correct role to create this.

2. Then, use SQL to create an external function on Snowflake—for example, named H2OPredict. The external function will reference the `api_integration`.

3. You are now ready to batch score a Snowflake table via an external function pass-through to an H2O eScorer. Here is a sample SQL statement:

```
select ID, H2OPredict('Modelname=riskmodel.zip',
loan_amnt, term, int_rate, installment, emp_length,
annual_inc, verification_status, addr_state, dti,
inq_last_6mths, revol_bal, revol_util, total_acc) as
H2OPrediction from RiskTable;
```

Let's have a look at Teradata integration.

Teradata integration

H2O.ai has partnered with Teradata to implement batch or real-time scoring directly against Teradata tables. This is done as shown in the following diagram:

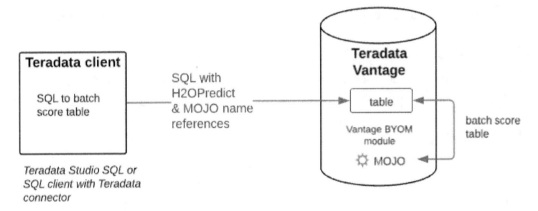

Figure 10.9 – MOJO scoring pattern for Teradata integration

Scoring context

Scoring is batch against Teradata tables and leverages the scalability of the Teradata platform. Thus, scoring can be made against any Teradata table, including those holding massive datasets. This is similar in concept to the Snowflake UDF integration, but only in concept: the underlying architectures and implementations are fundamentally different.

Scoring against Teradata tables is done by running a SQL statement from a Teradata client. This can be either a native Teradata Studio client or a SQL client with a Teradata connector.

Implementation

To implement, you first must install Teradata Vantage **Bring Your Own Model** (**BYOM**). Then, you use SQL to create a Vantage table to store H2O MOJOs. You then use SQL to load MOJOs into the Vantage table. Details can be found at `https://docs.` `teradata.com/r/CYNuZkahMT3u2Q~mX35YxA/WC6Ku8fmrVnx4cmPEqYoXA`.

Scoring example

Here is an example SQL statement to batch score a Teradata table:

```
select * from H2OPredict(
on risk_table
on (select * from mojo_models where model_id=riskmodel)
dimension
using Accumulate('id')
) as td_alias;
```

In this case, the code assumes all `risk_table` fields are used as input into the MOJO.

Additional notes

Your SQL statement to batch score may include options to return reason codes, stage probabilities, and leaf-node assignments.

BI tool integration

A powerful use of MOJO scoring is to integrate into BI tools. The most common way is to implement MOJO scoring either on a REST server or against a database, as shown in the following diagram. Note that in this pattern, MOJOs are not deployed on the BI tool itself, but rather, the tool integrates with an external scoring system. The low-latency nature of MOJO scoring allows users to interact in real time with MOJO predictions through this pattern:

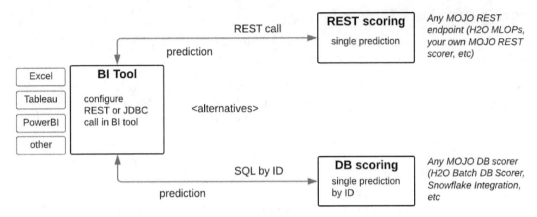

Figure 10.10 – MOJO scoring patterns for BI tool integration

Scoring context

BI tools integrate real-time predictions from external scorers.

Implementation

An external REST or database MOJO scoring system is implemented. Integration with the external scorer is implemented in the BI tool. These integrations are specific to each BI tool, and often, a single BI tool has multiple ways to make this integration.

Scoring example – Excel

The following code block shows a formula created in a cell of an Excel spreadsheet:

```
=WEBSERVICE(CONCAT("http://192.1.1.1:8080/
modeltext?name=riskmodel.mojo&row=",TEXTJOIN(",",FALSE,
$A4:$M4))))
```

This web service is called when the formula is applied to the target cell, or whenever a value changes in any of the cells referenced in the formula. A user can then drag the formula down a column and have predictions fill the column.

Note in the preceding formula that the REST call composes the observation to be scored as CSV and not as JSON. The structuring of this payload is specific to the REST API and its endpoint.

We can integrate MOJO scoring into other third-party software using REST endpoints in a similar fashion, though the semantics of the endpoint construction differ. Let's see how to do it in Tableau.

Scoring example – Tableau

Tableau is a common dashboarding tool used within enterprises to present information to a variety of different users within the organization.

Using the Tableau script syntax, a model can be invoked from the dashboard. This is very powerful as now, a business user can get current prediction results directly in the dashboard on demand. You can see an example script here:

```
SCRIPT_STR(
'name'='riskmodel.mojo',
ATTR([./riskmodel.mojo]),
ATTR([0a4bbd12-dcad-11ea-ab05-024200eg007]),
ATTR([loan_amnt]),
ATTR([term]),
ATTR([int_rate]),
ATTR([installment]),
ATTR([emp_length]),
ATTR([annual_inc]),
ATTR([verification_status]),
ATTR([addr_state]),
ATTR([dti]),
ATTR([inq_last_6mths]),
ATTR([revol_bal]),
ATTR([revol_util]),
ATTR([total_acc]))
```

The script reads the values as attributes (ATTR keyword) and passes them to a script in the Tableau environment when a REST call is made to the model. Using the REST call allows a centralized model to be deployed and managed, but different applications and consumers invoke the model based on their specific needs.

Now, let's see how to build a REST endpoint in Power BI.

Scoring example – Power BI

Here is a scoring example for Power BI. In this case, we are using a `Web.Contents` Power Query M function. This function is pasted to the desired Power BI element in your Power BI dashboard:

```
Web.Contents(
    "http://192.1.1.1:8080",
    [
        RelativePath="modeltext",
        Query=
        [
            name="riskmodel.mojo",
            loan_amnt=Loan_Ammt,
            term=Term,
            int_rate=Int_Rate,
            installment=Installments,
            emp_length=Emp_Length,
            annual_inc=Annual_Inc,
            verification_status=Verification_Status,
            addr_state=Addr_State,
            dti=DTI,
            inq_last_6mths= Inq_Last_6mths,
            revol_bal=Revol_Bal,
            revol_util=Revol_Util,
            total_acc=Total_Acc
        ]
    ]
)
```

Let's generalize a bit from these specific examples.

Additional notes

Each BI tool integrates with a REST endpoint or database in its own way and often provides multiple ways to do so. See your BI tool documentation for details.

UiPath integration

UiPath is an RPA platform that automates workflows based on human actions. Making predictions and responding to these predictions is a powerful part of this automation, and thus scoring models during these workflow steps is a perfect fit. You can see an example of this in the following diagram:

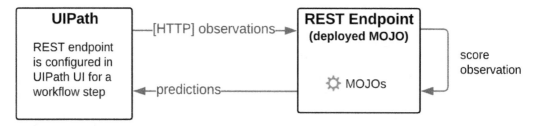

Figure 10.11 – MOJO scoring pattern for UiPath integration

Scoring context

UiPath integrates with external MOJO scoring similar to what was shown for BI tools in the previous section. In the case of UiPath, a workflow step is configured to make a REST call, receive a prediction, and respond to that prediction.

Implementation

MOJO scoring is implemented externally on a REST server, and the UiPath Request Builder wizard is used to configure a REST endpoint to return a prediction. Details can be seen here: `https://www.uipath.com/learning/video-tutorials/application-integration-rest-web-service-json`.

Scoring example

This video shows how to automate a workflow using H2O MOJO scoring: `https://www.youtube.com/watch?v=LRlGjphraTY`.

We have just finished our survey of some MOJO scoring patterns for third-party software. Let's look at a few scoring patterns with software that your organization builds itself.

Exploring examples of MOJO scoring with your target-system software

In addition to deploying MOJOs for scoring on H2O and third-party software, you can also take a **Do-It-Yourself** (**DIY**) approach and deploy scoring in your own software. Let's see how to do this.

Your software application

There are two ways to score from your own software: integrate with an external scoring system or embed scoring directly in your software system.

The following diagram shows the pattern of integrating with an external scoring system:

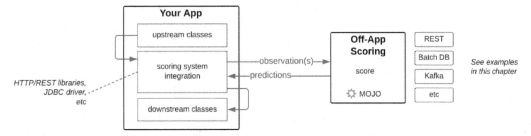

Figure 10.12 – MOJO application-scoring pattern for external scoring

This pattern should look familiar because it is fundamentally the same as what we saw with scoring from BI tools: your software acts as a client to consume MOJO predictions made from another system. The external prediction system can be a MOJO deployed on a REST server (for example, an H2O REST scorer) or batch database scorer (for example, a Snowflake Java UDF or an H2O batch database scorer) or another external system, and your application needs to implement the libraries to connect to that system.

In contrast, the following diagram shows the pattern of embedding MOJO scoring directly into your application itself:

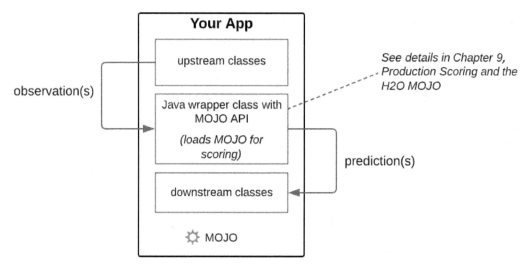

Figure 10.13 – MOJO application-scoring pattern for embedded scoring

Doing so requires your application to implement a Java wrapper class that uses the H2O MOJO API to load the MOJO and score data with it. This was shown in detail in *Chapter 9, Production Scoring and the H2O MOJO.*

When should you use the external versus embedded scoring pattern? There are, of course, advantages and disadvantages to each pattern.

The external scoring pattern decouples scoring from the application and thus allows each component and the personas around it to focus on what it does best. Application developers, for example, can focus on developing the application and not deploying and monitoring models. Additionally, an external scoring component can be reused so that many applications and clients can connect to the same deployed model. Finally, particularly in the case of on-database scoring (for example, Java UDF and Teradata integration) and streaming scoring with extreme batch size or throughput, it would be difficult or foolish to attempt to build this on your own.

The embedded scoring pattern has the advantage of eliminating the time cost of sending observations and predictions across the network. This may or may not be important depending on your **service-level agreements (SLAs)**. It certainly simplifies the infrastructure requirements to perform scoring, especially when network infrastructure is unavailable or unreliable. Finally, and often for regulatory reasons, it may be desirable or necessary to manage the model deployment and the application as a single entity, thus demanding the coupling of the two.

On-device scoring

MOJOs can be deployed to devices, whether they be an office scanner/printer, a medical device, or a sensor. These can be viewed as mini-applications, and the same decision for external or embedded scoring applies to devices as with applications, as discussed previously. In the case of devices, however, the advantages and disadvantages of external versus embedded scoring can be magnified greatly. For example, devices such as **internet of things (IoT)** sensors may number in the thousands, and the cost of deploying and managing models on each of these may outweigh the cost of greater latency resulting from network communication to a central external scorer.

> **Important Note**
> A rule of thumb is that the available device memory needs to be over two times the size of the MOJO.

Exploring examples of accelerators based on H2O Driverless AI integrations

This book thus far has focused on building models at scale using H2O. We have been doing this with H2O Core (often called H2O Open Source), a distributed ML framework that scales to massive datasets. We will see in *Chapter 13, Introducing H2O AI Cloud,* that H2O offers a broader set of capabilities represented by an end-to-end platform called H2O AI Cloud. One of these capabilities is a highly focused AI-based AutoML component called Driverless AI, and we will distinguish this from H2O Core in *Chapter 13, Introducing H2O AI Cloud.*

Driverless AI is like H2O Core because it also generates ready-to-deploy MOJOs with a generic MOJO runtime and API, though for Driverless AI a license file is required for MOJO deployment, and the MOJO and runtime are named differently than for H2O Core.

The reason for mentioning this here is that several integrations of Driverless AI have been built and are well documented but have not analogously been built for H2O Core. These integrations and their documentation can be used as accelerators to do the same for H2O Core. Just bear in mind the lack of license requirement for deploying H2O Core MOJOs, and the differently named MOJOs and runtime.

> **Approach to Describing Accelerators**
>
> Accelerators are overviewed here and links are provided to allow you to understand their implementation details. As noted, these accelerators represent the deployment of MOJOs generated from the H2O Driverless AI AutoML tool. Please review *Chapter 9, Production Scoring and the H2O MOJO* to understand how MOJOs generated from H2O Core (H2O-3 or Sparkling Water) are essentially the same as those generated from Driverless AI but with differences in naming and the MOJO API. This knowledge will allow you to implement the Driverless AI MOJO details shown in the links for H2O Core MOJOs.

Let's take a look at some examples.

Apache NiFi

Apache NiFi is an **open source software** (**OSS**) designed to program the flow of data in a UI and drag-and-drop fashion. It is built around the concept of moving data through different configurable processors that act on the data in specialized ways. The resulting data flows allow forking, merging, and nesting of sub-flows of processor sequences and generally resemble complex **directed acyclic graphs** (**DAGs**). The project's home page can be found here: `https://nifi.apache.org/index.html`.

NiFi processors can be used to communicate with external REST, JDBC, and Kafka systems and thus can leverage the pattern of scoring MOJOs from external systems.

You can, however, build your own processor that embeds the MOJO in the processor to score real-time or batch. This processor requires only configurations to point to the MOJO and its dependencies. The following link shows how to do this for Driverless AI and can be used as an accelerator for doing the same with H2O Core: `https://github.com/h2oai/dai-deployment-examples/tree/master/mojo-nifi`.

Apache Flink

Apache Flink is a high throughput distributed stream- and batch-processing engine with an extensive feature set to run event-driven, data analytics, and data pipeline applications in a fault-tolerant way.

The following link shows how to embed Driverless AI MOJOs to score data directly against Flink data streams and can be used as an accelerator for doing the same with H2O Core: `https://github.com/h2oai/dai-deployment-examples/tree/master/mojo-flink`.

AWS Lambda

AWS Lambda is a serverless computing service that lets you run code without the need to stand up, manage, and pay for underlying server infrastructure. It can perform any computing task that is short-lived and stateless, and thus is a nice fit for processing scoring requests. The following accelerator shows how to implement an AWS Lambda as a REST endpoint for real-time or batch MOJO scoring: `https://h2oai.github.io/dai-deployment-templates/aws_lambda_scorer/`.

AWS SageMaker

AWS SageMaker can be used to host and monitor model scoring. The following accelerator shows how to implement a REST endpoint for real-time MOJO scoring: `https://h2oai.github.io/dai-deployment-templates/aws-sagemaker-hosted-scorer/`.

And now, we have finished our survey of scoring and deployment patterns for H2O MOJOs. The business value of your H2O-at-scale models is achieved when they are deployed to production systems. The examples shown here are just a few possibilities, but they should give you an idea of how diverse MOJO deployments and scoring can be.

Let's summarize what we've learned in this chapter.

Summary

In this chapter, we explored a wide diversity of ways to deploy MOJOs and consume predictions. This included scoring against real-time, batch, and streaming data and scoring with H2O software, third-party software (such as BI tools and Snowflake tables), and your own software and devices. It should be evident from these examples that the H2O model-deployment possibilities are extremely diverse and therefore able to fit your specific scoring needs.

Now that we have learned how to deploy H2O models to production-scoring environments, let's take a step back and start seeing through the eyes of enterprise stakeholders who participate in all the steps needed to achieve success with ML at scale with H2O. In the next section, we will view H2O at scale through the needs and concerns of these stakeholders.

Section 4 – Enterprise Stakeholder Perspectives

In this section, we put on our enterprise stakeholder hats and learn how to plan for, deploy, administer, maintain, and secure the H2O at scale platform. You will first understand how Enterprise Steam works to integrate H2O with your enterprise server cluster and security environment. You will then learn how to configure Enterprise Steam to govern users and define the size of their H2O model-building environments, thereby controlling resource consumption and cost on your enterprise cluster. From there, we learn key areas of operations to support model building and deployment with H2O. Finally, we learn the details of H2O architecture and security from multiple architecture views. Data scientists are shown how these activities relate to their own needs.

This section comprises the following chapters:

- *Chapter 11, The Administrator and Operations Views*
- *Chapter 12, The Enterprise Architect and Security Views*

11
The Administrator and Operations Views

We have spent a good proportion of time in this book so far understanding the components of the H2O machine learning at scale framework and deep diving to develop the skills to implement the framework for model building and model deployment on enterprise systems.

Success in machine learning requires skill with code and technology but in an enterprise, it also requires success with people and processes. Looked at in another way, the proverbial *success is a team sport* statement is all too true. Let's now begin looking at the personas or stakeholders that participate in bringing success to H2O machine learning at scale.

In this chapter, we will start by addressing the personas directly involved in this success: H2O administrators, the operations team, and the data scientist. We'll understand how the key stakeholders interact with H2O at scale model building and model deployment. We'll look at the H2O administrator view of H2O at scale and understand how Enterprise Steam is central to this view. We'll also look at the operations team view of H2O at scale model building and model deployment. Finally, we'll understand how data scientists are impacted by the H2O administrator and the operations team views.

In this chapter, we're going to cover the following main topics:

- A model building and deployment view: The personas on the ground
- View 1: Enterprise Steam administrator
- View 2: The operations team
- View 3: The data scientist

A model building and deployment view – the personas on the ground

Many personas or stakeholders are involved in the machine learning life cycle. The key personas involved in H2O at scale model building and deployment are the data scientist, the Enterprise Steam administrator, and the operations team. The focus of this chapter is on these personas. Let's get a high-level view of their activities as shown in the following diagram:

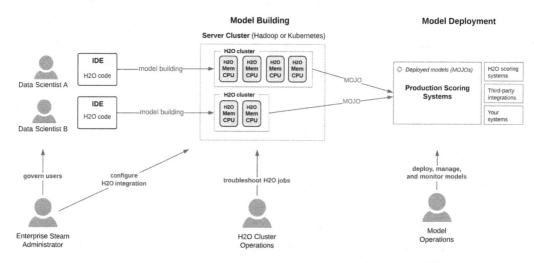

Figure 11.1 – Key personas involved in building and deploying H2O models at scale

A summary of these stakeholders and their high-level concerns are as follows:

- **Enterprise Steam Administrator**: Governs who launches H2O clusters on the multitenant enterprise server cluster and how many resources they are allowed to consume, and centralizes and manages H2O integration on the server cluster

- **H2O Cluster Operations**: Troubleshoots H2O jobs on the enterprise server cluster (either a Kubernetes or Hadoop cluster)

- **Model Operations**: Deploys, manages, and monitors models deployed to scoring environments

- **Data Scientist**: For the purpose of this chapter, knows how and why to interact with the aforementioned personas

The rest of this chapter will be dedicated to drilling down into each of these persona's views and activities when working on H2O machine learning at scale.

Let's get started.

View 1 – Enterprise Steam administrator

Let's first look at the concerns of the Enterprise Steam administrator before understanding their activities.

Enterprise Steam administrator concerns

The Enterprise Steam administrator has two broad concerns as shown in the following diagram:

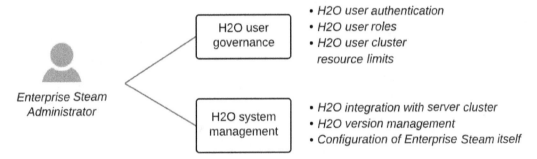

Figure 11.2 – Enterprise Steam administrator concerns

The preceding diagram can be summarized as follows:

- **H2O user governance**: These concerns focus on governing data scientists, who launch H2O model building jobs on the enterprise server cluster. This includes authenticating these H2O users against the enterprise identity provider, defining roles for H2O users, and governing the resource consumption of H2O users in the enterprise server cluster environment.

- **H2O system management**: These concerns revolve around centralizing the integration of H2O technology with the enterprise server cluster and managing H2O software versions. This creates a separation of roles where now data scientists do not need to understand the complexities of this integration (which they would if Enterprise Steam were not implemented).

Enterprise Steam can be seen as a necessity in handling enterprise concerns.

Enterprise Steam's Value to the Enterprise

Enterprises tend to be careful in governing who uses enterprise systems, how users integrate with them, and how users use resources on them. Administrators use Enterprise Steam to build centralized safeguards to handle these concerns and to create uniformity and predictability of H2O usage. Enterprise Steam also makes it easier for data scientists because they do not need to know the technical complexities of integrating H2O with the enterprise system.

To appreciate the benefits of Enterprise Steam, it is useful to look at the role it plays in the workflow of data scientists building models at scale with H2O. This is summarized in the following diagram:

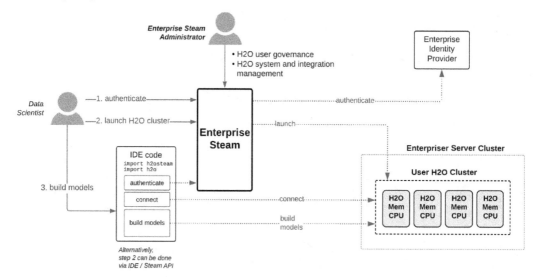

Figure 11.3 – Data science and administrator workflows through Enterprise Steam

Fundamentally, Enterprise Steam sits in front of all data scientists' access to H2O on the enterprise server cluster. Let's understand the benefits of this design in a step-by-step way:

1. **Data scientist authenticates through Enterprise Steam**: All data scientists using H2O are authenticated through the enterprise identity provider, for example, LDAP. We will see later in the chapter that group membership returned from the identity provider is used to configure user capabilities specific to the group.

2. **Launch H2O cluster**: The data scientist must go through Enterprise Steam to launch an H2O cluster on the enterprise server cluster. In doing so, the data scientist is given boundaries around the size of the H2O cluster (the number of server nodes, memory, and CPU per node) that can be launched and how long it can run before being shut down. We will see later in the chapter how the Enterprise Steam admin uses a **profile** to define these boundaries.

    Note that there is a lot of complexity involved in configuring H2O to integrate with the specifics of the enterprise server cluster. This is taken care of by the Enterprise Steam administrator using Enterprise Steam's configuration workflow. Data scientists are shielded from this task and simply use the Enterprise Steam UI (or its API from the model building IDE) to quickly launch an H2O cluster.

3. **Build models**: From the model building IDE, data scientists must use the Enterprise Steam API to authenticate before interacting with the H2O cluster that has been launched. After authenticating through Enterprise Steam, the data scientist will be able to connect to their H2O cluster and write code to build models on it.

Note that all the steps in the workflow just described can be done programmatically from the data scientist's IDE and do not require interaction with the Enterprise Steam UI.

The Enterprise Steam UI versus the API

It is convenient to manage one or more H2O clusters (launch, stop, re-launch, terminate) from the Enterprise Steam UI. Alternatively, this can be done from your IDE as well by using the same Enterprise Steam API required to authenticate to Enterprise Steam when connecting to your cluster from the IDE to build models. Thus, you can work entirely from your IDE if you wish.

Let's now return to the role of the Enterprise Steam administrator in governing H2O users and managing the H2O system and integration.

Enterprise Steam configurations

Let's go directly to the home configurations screen in Enterprise Steam. This will help us get the big picture before focusing on H2O user governance and H2O system management. This is shown in the following screenshot. Keep in mind that only Enterprise Steam administrators can see and access these configurations:

Home > Configurations

CONFIGURATIONS

System-wide configuration for Enterprise Steam. Please refer to the documentation ⌐. All setting are validated before saving. Some settings require Enterprise Steam to be restarted to take effect.

GENERAL

ACCESS CONTROL					STEAM CONFIGURATION			
Authentication	Token	Users	Roles	Profiles	Licensing	Security	Logging	Import/Export

BACKENDS

HADOOP	disabled	KUBERNETES	enabled
Configuration		Configuration	
		Volume Mounts	
		HDFS and Hive	
		Minio	
		H2O.ai Storage	

PRODUCTS

H2O	running	SPARKLING WATER	disabled	DRIVERLESS AI	disabled
Configuration		Configuration		Configuration	
Engines		Engines		Engines	
Startup Parameters		Python Environments		Python Client	

Figure 11.4 – Enterprise Steam configurations home page (administrator visibility only)

Let's organize this logically:

- **H2O user governance**: This is done using the **ACCESS CONTROL** configuration set. Note that these configurations also include user resource consumption governance and other guardrails.

- **H2O System Management – integration**: Configurations for integration with the enterprise cluster environment are done using the **BACKENDS** and **PRODUCTS** configuration sets. This includes management of H2O versions as well.

- **H2O System Management – Steam**: Configuring Enterprise Steam itself is done using the **STEAM CONFIGURATION** configuration set.

> **Driverless AI Can be Configured in Enterprise Steam (Wait, What?)**
>
> The focus of this book is machine learning at scale with H2O. From a model building perspective, this focuses on H2O-3 or Sparkling Water clusters training on massive datasets via the horizontally scaling architecture these create.
>
> Driverless AI is an H2O product that is specialized for extreme AutoML (typically on datasets of less than 100-200 GB) leveraging extensive automation, a genetic algorithm to find the best models, and highly automated and exhaustive feature engineering. We will elaborate more on Driverless AI when covering H2O.ai's end-to-end machine learning platform in *Chapter 13, Introducing H2O AI Cloud*. We will see in that chapter how Driverless AI can augment the use of H2O-3 and Sparkling Water at scale. For now, know that Driverless AI can also be launched and managed through Steam (on a Kubernetes cluster).

Let's now elaborate on the previous bullets one by one.

H2O user governance from Enterprise Steam

Enterprise Steam administrators govern users of H2O clusters using the **GENERAL > ACCESS CONTROL** set of configurations. An overview of these configurations follows.

Authentication

User authentication to Enterprise Steam and thus to H2O clusters that are launched can be configured against the following enterprise identity providers:

- **OpenID**
- **LDAP**
- **SAML**
- **PAM**

Detailed and familiar settings are configured depending on which provider is selected.

Tokens

Tokens are alternatives to using passwords when using the Enterprise Steam API. Tokens are issued here for the logged-in user. Each time a token is generated the previously issued token is revoked. In the case of OIDC auth, the user must obtain a token to use the API (a **single sign-on (SSO)** password cannot be used).

Users

Users who have authenticated through Enterprise Steam are listed on this page. Users are listed by username, role, and authentication method. Users can be deactivated and reactivated from here.

Overrides of an individual user's role, authentication method, and profile assignments can be done here as well. Note that the user's role, authentication method, and profile are typically assigned via mapping to a group that they belong to and that is returned by the identity provider. Also note that a user configuration for enabled identity provider can be overridden here when the user exists in more than one enabled identity provider system.

Roles

There are two roles in Enterprise steam:

- **Admin**: Logged-in users with admin roles can make configuration changes as discussed in this chapter.

- **Standard user**: Logged-in users with standard user roles are data scientists who can launch clusters from Enterprise Steam as shown in *Figure 11.3*. This user experience is described more fully in *Chapter 3, Fundamental Workflow – Data to Deployable Model*.

Note that roles can be assigned by group name (returned from authentication against the identity provider). You may also provide a group name with the wildcard character * to assign a role to all authenticated users.

Profiles

This is where users are given boundaries on the resource consumption of their H2O clusters. These boundaries are assigned to a profile that is given a name, and profiles are mapped to one or more user groups or to individual users. The wildcard character * maps a profile to all users.

Users are constrained to these boundaries when they launch an H2O cluster. For example, an intern may only be allowed one concurrent H2O cluster comprising no more than 2 nodes each with 1 GB of memory, whereas an advanced user may be allowed 3 concurrent clusters each with up to 20 nodes and 50 GB of memory per node.

> **Profiles Govern User Resource Consumption**
>
> Profiles draw boundaries around each user's resource consumption on the shared server cluster. This is done by limiting the number of H2O clusters a user can manage concurrently, the size of each cluster (total server nodes, memory per node, CPU per node), and how long the cluster can run.
>
> Profiles should not limit users but on the other hand, they should be appropriately sized for them. Without profiles, users tend toward maximum possible resource consumption while requiring far less. Multiplied by all users this typically creates unnecessary pressure to expand the server cluster (which has a significant cost) or to constrain the tenants that use it.

There are three profile types for H2O Core that can be configured and assigned to users:

- **H2O**: H2O-3 clusters on Hadoop.

- **Sparkling Water – Internal Backend**: Sparkling Water clusters on Hadoop where H2O and Spark DataFrames occupy the same memory space (that is, memory on the same server node).

- **Hadoop: Sparkling Water – External Backend**: Sparkling Water clusters on Hadoop where H2O and Spark DataFrames occupy separate memory space (memory on different server nodes). Note that this profile is used infrequently, typically only for Sparkling Water processing that lasts extremely long durations (typically over 24 hours). The external backend has the advantage of isolating H2O Sparkling Water clusters from disruption caused by Spark node termination or reassignment but has the disadvantage of requiring network (versus memory space) transfer of data between H2O and Spark DataFrames.

- **H2O – Kubernetes**: H2O-3 clusters on the Kubernetes cluster framework.

See *Chapter 2, Platform Components and Key Concepts*, to revisit the distinction between H2O-3 and Sparkling Water clusters. Also recall that **H2O cluster** refers to either an **H2O-3 cluster** or a **Sparkling Water cluster**.

> **Hadoop Spark versus a Pure Spark Cluster**
>
> Hadoop systems typically implement both MapReduce and Spark frameworks as distributed compute systems and jobs on these frameworks typically are managed by the YARN resource manager. Enterprise Steam manages H2O-3 clusters on the MapReduce framework of Hadoop/YARN and Sparkling Water clusters on the Spark framework of Hadoop/YARN.
>
> Note that Spark can also be run alone outside of Hadoop and YARN. H2O Sparkling Water clusters can be run on this type of Spark implementation but currently, H2O does not integrate Enterprise Steam with these environments.
>
> Kubernetes is an alternative framework for implementing distributed compute. Enterprise Steam also integrates with Kubernetes to launch and run H2O-3 clusters. Support for Enterprise Steam and Kubernetes for Sparkling Water is currently in progress.

Details differ for the four profile types listed previously, but they each share the following key configurations:

- **Profile name**: To assign to user groups.
- **User groups**: Who to assign the profile to.
- **Cluster limit per user**: The number of H2O clusters a user can concurrently run.
- **Number of nodes**: The number of server nodes comprising the distributed H2O cluster.
- **Memory and CPU**: The amount of memory and CPU per node to dedicate to the H2O cluster.
- **Maximum idle time (hrs)**: The H2O cluster will automatically shut down when the H2O cluster is idle for longer than this duration.
- **Maximum uptime (hrs)**: The H2O cluster will automatically shut down when the H2O cluster has been running longer than this duration, whether idle or not.
- **Enable cluster saving**: Saves cluster data when the cluster is shut down (either after the user manually shuts it down or after the maximum idle time or uptime is exceeded) so that the cluster can be restarted and resume with the same state as when it shut down.

Note that the preceding is not an exhaustive list but just a list of the key configurations to govern user resource usage on an H2O cluster.

Enterprise Steam configurations

Configurations to manage the Enterprise Steam server are made at **GENERAL** > **STEAM CONFIGURATION**. The configuration sets to do this are as follows.

License

Enterprise Steam requires a license from H2O.ai. This configuration page identifies the number of days remaining on the current license and provides ways to manage the license.

Security

This configuration page provides settings to harden the security of Enterprise Steam. Enterprise Steam only runs on HTTPS and Steam by default generates a self-signed certificate. This page allows you to configure the path to your own TLS certificate, among other security settings.

Logging

Enterprise Steam performs extensive logging of user authentication and H2O cluster usage. This page allows you to configure the log level and the log directory path. This is also where you can download the logs.

You can also download usage reports of H2O clusters at user-level granularity.

Import/export

This page facilitates the reuse of configurations by allowing the Enterprise Steam configuration to be exported as well as a configuration from another instance to be imported and loaded.

Server cluster (backend) integration

A theme of this book is that H2O Core (H2O-3 and Sparkling Water) is used to build models on massive data volumes. To achieve this, users launch H2O clusters, which distribute both data and compute in parallel across multiple separate servers. This allows, for example, XGBoost or **Generalized Linear Model** (GLM) algorithms to train against a terabyte of data.

Enterprise Steam allows the administrators to configure the integration of H2O clusters in these server cluster environments. This is done in the **BACKENDS** section of the configuration pages.

There are two types of server cluster backends that H2O clusters can be launched on and governed by Enterprise Steam:

- **Hadoop**: These are YARN-based distributed systems, for example, Cloudera CDH or CDP, or Amazon EMR.

- **Kubernetes**: This is a distributed framework architected around orchestrating pools of containers and is vendor-agnostic, though with vendor-specific offerings.

Configuration items are detailed and extensive for either backend to allow H2O clusters to run securely against these enterprise environments. See the H2O documentation for full details at `https://docs.h2o.ai/enterprise-steam/latest-stable/docs/install-docs/backends.html`.

H2O-3 and Sparkling Water management

Now that cluster backends have been configured, administrators can manage and configure **H2O-3** or **Sparkling Water**, which users run on these backends. This is done in the **PRODUCTS** section of the configuration pages.

Both H2O-3 and Sparkling Water have the following pages:

- **Configuration**: This presents high-level configurations that will be constant for all H2O-3 or Sparkling Water clusters among users and profiles. For example, you can append a configured YARN prefix to the job name appearing for all H2O clusters listed in the YARN resource manager UI on the Hadoop system.

- **Engines**: Engines refers to H2O-3 or Sparkling Water library versions that are used to launch an H2O cluster. This configuration page lists all library versions (engines) maintained by Enterprise Steam. Each is available as a drop-down choice when the user launches an H2O cluster. Note that new library versions are either uploaded to the library (after first downloading from the H2O.ai website) or copied to the Enterprise Steam server local filesystem (for example, via the `scp` command from another server). Library versions can be added and removed on this configuration page.

> **Note on Upgrading H2O Versions**
>
> Upgrading H2O versions is easy: the user simply launches a new H2O cluster by selecting a new library version (that the administrator has uploaded or copied to Enterprise Steam as just described). This simple upgrade method works because (a) the H2O cluster architecture pushes the library from Enterprise Steam to the nodes on the server cluster when the H2O cluster is launched and removes the library when the H2O cluster is terminated, and (b) each H2O cluster is an isolated entity. Thus, *user A* can launch an H2O cluster using one H2O version and *user B* can do so using another version (or *user A* can launch multiple H2O clusters each with a different version).
>
> A requirement in all cases is that the library version installed in your IDE environment matches the version the H2O cluster was launched with.

See the H2O.ai documentation for more in-depth details about using H2O-3 on Hadoop: `https://docs.h2o.ai/h2o/latest-stable/h2o-docs/welcome.html#hadoop-users`. Note that the list of Hadoop launch parameters listed at this link can be used to configure Enterprise Steam for all H2O-3 clusters.

Restarting Enterprise Steam

For some configuration changes, you will be prompted with a message stating that a restart of Enterprise Steam is necessary to apply these changes.

Restart Enterprise Steam by running the following command from the Enterprise Steam server command line:

```
sudo systemctl restart steam
```

Validate that Enterprise Steam is running after the restart by running the following command:

```
sudo systemctl status steam
```

The log for troubleshooting the Enterprise Steam service is found at the following path on the Enterprise Steam server: `/opt/h2oai/steam/logs/steam.log`.

Now that we have understood the Enterprise Steam administrator views of machine learning at scale with H2O, let's see what this means to the operations team.

View 2 – The operations team

The operations team maintains enterprise systems and monitors the workloads run on them. For H2O at scale, these operations focus on the following three areas:

- The Enterprise Steam server
- H2O model building jobs running on the enterprise cluster
- The models deployed to production scoring environments

Specific operations personas around these areas and their concerns are summarized in the following diagram:

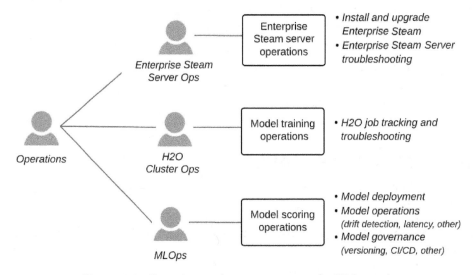

Figure 11.5 – Operations and persona concerns for H2O at scale

Let's look more closely at each operations persona and their concerns around H2O at scale.

Enterprise Steam server Ops

Ops around the Enterprise Steam server are focused on deploying and maintaining the server and the Enterprise Steam service that runs on it.

Enterprise Steam is a lightweight service that is mostly a web app with an embedded database to maintain state. There are no data science workloads that run on Enterprise Steam and no data science data passes through or resides on it. As discussed previously and summarized in *Figure 11.3*, Enterprise Steam launches the H2O cluster formation on the backend Hadoop or Kubernetes infrastructure where H2O workloads occur as driven by the data scientist's IDE.

H2O cluster Ops

H2O cluster ops are typically performed by Hadoop or Kubernetes administrators. On Hadoop, H2O clusters are run as native YARN MapReduce (for H2O-3 clusters) or Spark (for Sparkling Water clusters) jobs and are viewable as such in the YARN resource management UI. As noted in *Chapter 2, Platform Components and Key Concepts*, each YARN job maps to a single H2O cluster and runs for the duration that the cluster runs for.

MLOps

MLOps sometimes refers to operations on the full machine learning life cycle but we will focus on operations around deploying and maintaining models for scoring in production systems. These operations typically focus on the following:

- **Model deployment**: This concerns the actual deployment of the model to its scoring environment and is typically performed via automated **continuous integration and continuous delivery (CI/CD)** pipelines.

- **Model monitoring (software focus)**: This is a traditional concern around asking how healthy the model is from a running software perspective. Typically, the model is monitored to see that it is running with no errors and that its scoring latencies are meeting **service-level agreement (SLA)** expectations.

- **Model monitoring (machine learning focus)**: This is a set of concerns around model scoring outcomes specifically. It typically addresses whether data drift is occurring (whether the data distribution at scoring time has shifted from that of the training data), which is used to determine whether the model needs to be retrained or not. Other concerns may be monitoring for prediction decay (whether the prediction distribution of scores is shifting toward being less predictive), monitoring for bias (whether a model is predicting in an unfair or prejudiced way for a particular subset or demographic group), and monitoring for adversarial attacks (attempts to fool models with artificial data to create a mistaken prediction or to maliciously damage the model scoring).

- **Model governance**: This is a set of concerns that include managing model versions and tracing deployed models back to the details of model building (for example, the training and testing datasets used to build the model, configuration details of the trained model, the data science owner of the trained model, and so on). It also includes the ability to roll back or reproduce models and other concerns that may be specific to the organization. Note that the model documentation generated during the model building stage is an excellent asset for use in model governance. (See *Chapter 7, Understanding ML Models*, for details on generating H2O AutoDoc.)

Note that MLOps is a rapidly moving set of practices and implementations. Building an MLOps framework on your own will likely result in great difficulty keeping up with the capabilities and ease of use of MLOps platforms offered by software vendors. We will see in *Chapter 13*, *Introducing H2O AI Cloud*, that H2O.ai offers a fully capable MLOps component as part of its end-to-end platform.

Now that we have understood the Enterprise Steam administrator and the operations views of machine learning at scale with H2O, let's see what this means to the data scientist.

View 3 – The data scientist

The primary ways that data scientists using H2O at scale interact with the Enterprise Steam administrator and the operations teams are shown in the following diagram:

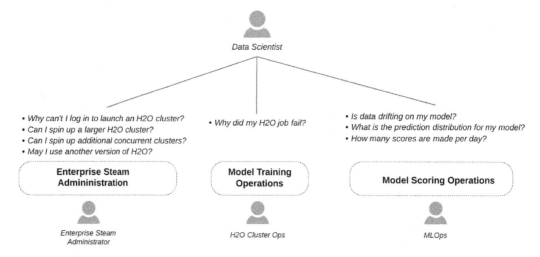

Figure 11.6 – Data scientist interactions with administrator and operations stakeholders when using H2O at scale

Let's drill down into the key data scientist interactions with the Enterprise Steam administrator and with operations teams.

Interactions with Enterprise Steam administrators

Recall that data scientists authenticate through Enterprise Steam and launch H2O cluster sizes as defined by Enterprise Steam profiles. Data scientists may at times interact with Enterprise Steam administrators to solve authentication issues or to ask for profiles that size H2O clusters more appropriately to their needs. They may also request access to a different YARN queue than what is allowed in their profile, request a longer configured idle time for their profile, or request a new version of H2O for launching H2O clusters. There may of course be other H2O configuration and profile requests. Reviewing the *View 1 – The Enterprise Steam administrator* section will help data scientists understand how Enterprise Steam configurations affect them and perhaps should be changed.

Interactions with H2O cluster (Hadoop or Kubernetes) Ops teams

Data scientists rarely interact with H2O cluster operations teams. These teams are typically Hadoop or Kubernetes operations teams, depending on which server cluster backend H2O clusters are launched on.

When data scientists do interact with these teams, it is typically to help troubleshoot H2O jobs that are failing or performing badly. This interaction may involve requests for Hadoop or Kubernetes logs to send to the **H2O support portal** for troubleshooting on the H2O side.

Interactions with MLOps teams

Data scientists typically interact with the MLOps team to engage at the beginning of the model deployment process. This often involves staging the model scoring artifact known as the H2O MOJO (generated from model building) for deployment and possibly staging other assets to archive with the model for governance purposes (for example, H2O AutoDoc).

Data scientists may also interact with the MLOps team to determine whether data drift is occurring to decide on whether to retrain the model with more up-to-date training data. Depending on the MLOps system, this process may be automated so that an alert is sent to the data scientist and other stakeholders with content that identifies and describes the drift.

Summary

In this chapter, we learned that H2O administrators (working through Enterprise Steam) and the operations team (managing the enterprise server cluster environment where H2O model building is executed, and also managing, monitoring, and governing the models after they are deployed to a scoring environment) are key personas participating in H2O machine learning at scale. We also learned how data scientists who build models are impacted by these personas and why they may need to interact with them.

Let's move on to two additional personas who play a role in H2O machine learning at scale and who may impact the data scientist: the enterprise architect and the security stakeholders.

12
The Enterprise Architect and Security Views

H2O at Scale is used to build state-of-the-art **machine learning** (**ML**) models against large-to-massive volumes of data on enterprise systems. Enterprise systems are complex integrations of diverse components that work together under common architecture and security principles. H2O at Scale needs to fit into this ecosystem in expected, secure, and cohesive ways. In this chapter, we will examine H2O at Scale architecture and security attributes in detail to understand how H2O software deploys to the enterprise, integrates with well-known frameworks, implements extensive security capabilities, and generally plays nicely in enterprise systems.

In this chapter, we will cover the following main topics:

- The enterprise and security architect view
- H2O at Scale enterprise architecture
- H2O at Scale security
- The data scientist's view of enterprise and security architecture

Technical requirements

There are no technical requirements for this chapter.

The enterprise and security architect view

Enterprise and security architects have broad roles, but for our purposes here, we will focus on their concerns about how software architecture integrates into and impacts the existing enterprise software ecosystem and how it addresses security requirements.

A summary of this view is shown in the following diagram. Enterprises will likely have their own specific needs and concerns, but our discussion here will be a common starting point:

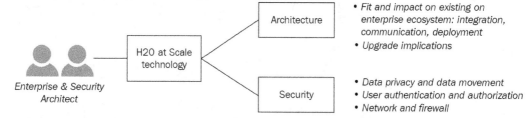

Figure 12.1 – Architecture and security views of H2O at Scale

Let's look at an overview before we dive into detailed views, as follows:

- **H2O enterprise architecture**: Needs and concerns here relate to how software components are deployed, integrated into, and potentially impact the rest of the technology ecosystem. Architecture in our case is primarily at the component level—that is, at the level of the *big separate pieces*—and not lower down, such as class design.

- **H2O security**: Enterprises go to extreme lengths to secure their systems from internal and external threats. H2O at Scale technology is used by a large proportion of *Fortune 100* and *Fortune 500* companies across industry verticals and has undergone and passed extensive security reviews to do so. We will touch on H2O security around the most scrutinized areas.

Let's first dive into the enterprise architecture view.

H2O at Scale enterprise architecture

We will look at H2O at Scale model building architecture through multiple lenses to allow architect stakeholders to understand how this technology addresses their common needs and concerns.

H2O at Scale components are flexible in how they are implemented, so first, we need to understand these alternatives and their implications, as elaborated in the following section.

H2O at Scale implementation patterns

There are three patterns for how H2O at Scale is implemented, as shown in the following diagram:

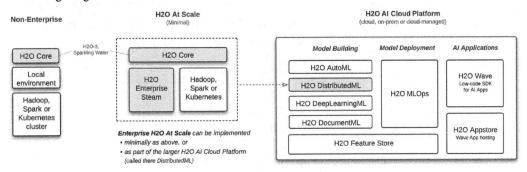

Figure 12.2 – Implementation patterns for H2O at Scale

A central concept to these patterns is that we are defining H2O at Scale components as Enterprise Steam, H2O Core (H2O-3 or Sparkling Water or both, with exportable H2O **MOJO** (short for **Model Object, Optimized**)), and an enterprise-server cluster (which H2O leverages but is not installed on). The reason for this component definition has been stated in many places throughout this book and will be recapped briefly here, as follows:

- **H2O at Scale (minimal)**: Minimally, H2O at Scale components are Enterprise Steam, H2O Core (H2O-3, Sparkling Water), and a Kubernetes or **Yet Another Resource Negotiator** (**YARN**)-based Hadoop or Spark server cluster. An H2O MOJO is generated from H2O Core. Its deployment to scoring systems was covered extensively in *Chapter 10, H2O Model Deployment Patterns*.

- **H2O AI Cloud platform**: H2O at Scale can be implemented as a subset of a larger H2O AI Cloud platform. This larger platform will be overviewed in *Chapter 13, Introducing H2O AI Cloud*. In this context, H2O at Scale retains its functionality and architecture and is considered one of multiple specialized model-building engines on the larger integrated H2O platform.

- **Non-enterprise H2O**: H2O Core can be implemented in two other patterns that are generally not considered enterprise implementations. H2O Core can be run on a user's local machine. This is useful for personal work, but the true power of H2O at Scale (in-memory data and its compute distributed horizontally across a server cluster) is not available in this case. Model building thus cannot be conducted at large data volumes.

  H2O Core can also be run directly against Hadoop, Spark, or Kubernetes clusters without implementing Enterprise Steam. This creates significant user management and governance risks for the enterprise and forces technical integration knowledge onto data scientists. This topic has been addressed in detail in *Chapter 11, The Administrator and Operations Views*, and will be touched on in this chapter.

> **Which H2O at Scale?**
>
> The architecture and security discussion for this chapter focuses on H2O at Scale components implemented minimally (the middle diagram in *Figure 12.2*). For H2O at Scale as a subset of the larger H2O AI Cloud platform, the architecture is similar but not identical. The differences will be outlined in *Chapter 13, Introducing H2O AI Cloud*.
>
> All discussions of H2O at Scale functionality, from model building to model deployment to Enterprise Steam administration, are the same whether H2O at Scale is implemented alone or as part of the larger H2O AI Cloud.

Let's start looking at how the user client, Enterprise Steam, and H2O Core components fit into the enterprise environment.

Component integration architecture

The following components are involved in H2O at Scale model building:

- A web browser
- A data-science **integrated development environment** (IDE) (for example, Jupyter Notebook) with an H2O client library installed
- H2O Enterprise Steam, with H2O-3 and/or Sparkling Water **Java ARchive** (JAR) files loaded for distributing as jobs to an enterprise-server cluster
- An enterprise **identity provider** (IdP)
- An enterprise-server cluster

These component integrations are shown in greater detail in the following diagram:

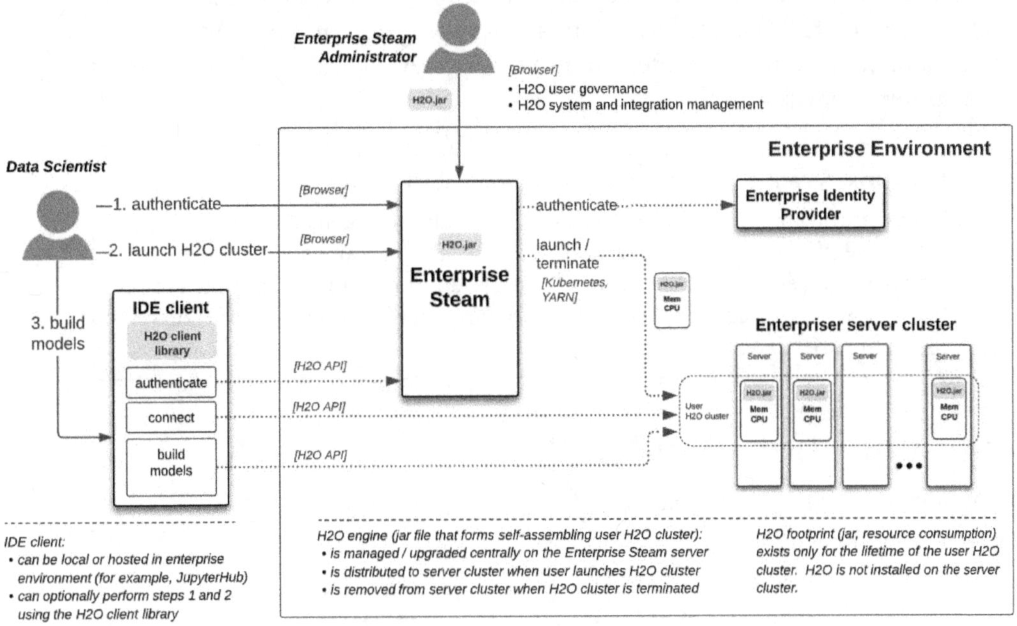

Figure 12.3 – Component integration architecture for H2O at Scale

Note that the only installed H2O components are an H2O and Enterprise Steam client library in the data scientist's IDE and H2O Enterprise Steam, which is installed on a small server. Enterprise Steam manages a Java JAR file that is pushed to the enterprise-server cluster to form a self-assembling H2O cluster for each user. Importantly, this JAR file is removed from the enterprise-server cluster when the H2O cluster is stopped. Thus, no H2O artifacts are installed on the enterprise-server cluster.

H2O at Scale Has a Small Component Footprint on the Enterprise System

Despite its ability to build models against massive datasets, the actual footprint of H2O at Scale on the enterprise system is quite small from an installed component standpoint.

Enterprise Steam is the only component installed in the enterprise environment and it requires only a small single server (an edge node when on Hadoop). The software to create H2O clusters (distributed model building) is managed on Enterprise Steam and pushed to the enterprise-server cluster temporarily and not installed there. Enterprise Steam leverages Kubernetes or YARN to launch an H2O cluster natively to these frameworks (that is, an H2O cluster is a native Kubernetes or YARN job).

Other than that, the data scientist installs an H2O client library in their IDE environment to communicate with Enterprise Steam and the user's H2O cluster.

Note also that all data scientist and administrator interactions with the enterprise environment are centralized through Enterprise Steam. This includes all user authentication and H2O cluster formation (and stopping or termination) and administrator configurations to integrate and secure H2O at Scale activities against the enterprise-server environment.

> **Enterprise Steam as a Centralized Component for H2O at Scale**
>
> Enterprise Steam serves as a central gateway for all user interactions with the enterprise system, including user authentication through an enterprise IdP, user management of the H2O cluster life cycle, administrator management of H2O software, and administrator configuration to integrate and secure H2O on the enterprise-server cluster.

Communication architecture

Communication protocols among H2O at Scale components are shown in the following diagram:

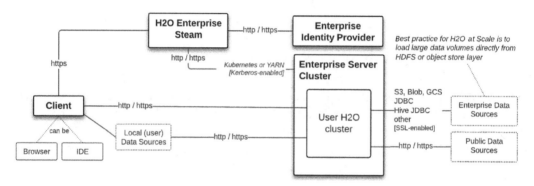

Figure 12.4 – Communication architecture of H2O at Scale

Note that H2O uses well-known communication protocols. Also, note that (except for data on the user's client machine) data is transferred directly from the data source to the user's H2O cluster (where it is partitioned into memory).

> **Data Ingest Is Direct from the Source (Preferably Storage Layer) to the H2O Cluster**
>
> Data ingested to the H2O cluster for model building does not pass through Enterprise Steam or the user's client (except if uploaded from the client): it passes directly from the data source to H2O's in-memory compute on the enterprise-server cluster. When dealing with massive datasets, it is best to use the server-cluster storage layer (for example, **Simple Storage Service (S3)**, Azure Blob, or **Hadoop Distributed File System (HDFS)**) as a data source compared to other data sources such as **Java Database Connectivity (JDBC)**. The H2O cluster partitions source data into memory among the nodes on the cluster. This loading of partitions is parallelized from the storage layer and not from other data sources (for example, a JDBC database or Hive).

Deployment architecture

There are two deployment patterns for H2O at Scale: on either Kubernetes- or YARN-based server clusters. In both cases (as discussed previously), only Enterprise Steam is installed on the enterprise environment. This Kubernetes- versus YARN-based distinction—and the architecture and infrastructure details that follow—are hidden from the data scientist. As stressed throughout this book, the data scientist uses familiar languages and IDEs to build H2O models at scale and knows little or nothing of the technical implementation of scaling that takes place behind the scenes on the server clusters.

Deployment on Kubernetes clusters

Kubernetes is a modern framework to scale and manage containerized applications. H2O at Scale can be deployed to leverage this framework natively. Details of how this is done are shown in the following deployment diagram:

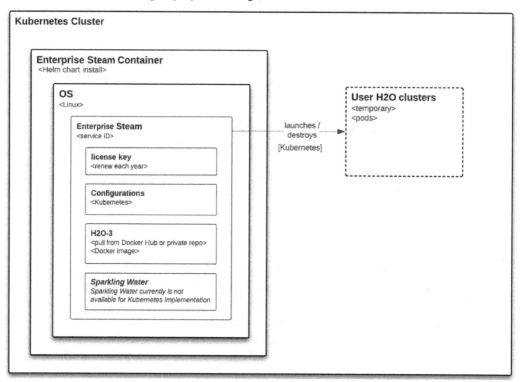

Figure 12.5 – Deployment diagram of H2O at Scale on Kubernetes

The key points are noted here:

- Enterprise Steam is installed and run on the Kubernetes cluster using a **Helm chart**. Helm is a package manager for Kubernetes, and Helm charts are a collection of templatized files in a directory structure used to define, install, and upgrade the components on the Kubernetes framework. Helm charts make installation and upgrades on Kubernetes easy and repeatable.

- Enterprise Steam is fundamentally a web-server application that pushes workloads to the enterprise cluster as an orchestrated **Kubernetes Pod**. As such, Enterprise Steam requires very low resources.

- Enterprise Steam runs as a service with a **service identifier (service ID)**.

- A **license key** acquired from H2O.ai is required to use Enterprise Steam. The license key expires in 1 year and needs to be renewed before this time.

- Kubernetes configurations are made in Enterprise Steam. These include **volume mounts**, **HDFS** and **Hive** access, and MinIO access (which opens access to the cloud object store—specifically, **Amazon Web Services (AWS) S3**, **Azure Blob**, and **Google Cloud Storage (GCS)**).

- H2O-3 **Docker images** are stored by version in any public (for example, **Docker Hub**) or private (for example, **AWS Elastic Container Registry (ECR)** image repository and are pulled from here by Kubernetes when the cluster is launched.

> **Note**
>
> H2O Sparkling Water is currently not available on Enterprise Steam on Kubernetes.

- For details on installing Enterprise Steam, see the H2O documentation at `https://docs.h2o.ai/enterprise-steam/latest-stable/docs/install-docs/installation.html`.

> **Upgrading H2O-3 is Easy**
>
> Recall that H2O-3 is not installed on the enterprise-server cluster: here, it is orchestrated and distributed as Pods on the Kubernetes cluster and then torn down after the user is done using the resulting H2O cluster. This makes upgrading easy: the administrator simply configures a new version of H2O in Enterprise Steam (and stores its Docker image in the configured repository), and users launch their next H2O cluster with the newer versions. (Multiple versions can coexist on Enterprise Steam, so users can select among them when launching an H2O cluster.)

For **high availability**, Enterprise Steam can be installed as an active-passive setup with automatic failover handled by Kubernetes. This is described in the H2O documentation at `https://docs.h2o.ai/enterprise-steam/latest-stable/docs/install-docs/ha.html`.

Let's see how H2O at Scale is deployed on YARN-based clusters. Fundamentally, these modes are similar when ignoring the underlying framework specifics.

Deployment on YARN-based server clusters

YARN is an open source framework for launching and managing distributed processing in a horizontally scalable architecture. It is typically implemented against Hadoop or Spark clusters (or Spark on Hadoop). The diagram and discussion that follows relate to deployments on Hadoop:

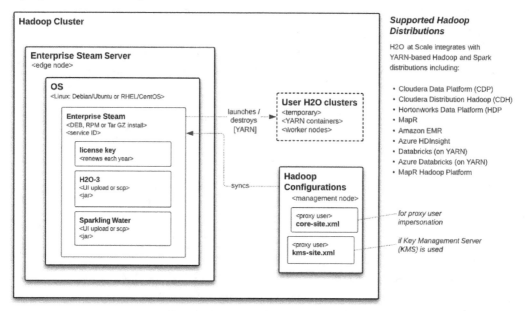

Figure 12.6 – Deployment diagram of H2O at Scale on Hadoop

Key points are similar to Kubernetes-based implementations. Differences for Hadoop are noted here:

- Enterprise Steam is deployed on a Hadoop **edge node**.

- Enterprise Steam pushes workloads to the enterprise cluster as a native **YARN** job.

- Enterprise Steam is installed on a **Linux OS** (either **Debian/Ubuntu** or **Red Hat Enterprise Linux (RHEL)/CentOS**) as a package (**Debian (DEB)** or **RPM Package Manager (RPM)** or a **tape archive GNU zip (TAR GZ)** installation).

- Configurations are made on the Hadoop **management node** to the core-site.xml and kms-site.xml configuration files. These Hadoop configurations are made to integrate Steam with YARN and to implement **user impersonation**. Because Enterprise Steam is on an edge node, changes to Hadoop configurations on the management node are replicated immediately on Enterprise Steam.

- H2O-3 or Sparkling Water JAR files are packaged by Enterprise Steam **YARN applications** and run against the Hadoop cluster as a separate YARN job for each user launching an H2O cluster. The YARN job (or, equivalently, the H2O cluster) runs until the user stops the H2O cluster, or Enterprise Steam auto-stops the cluster after a threshold of idle or absolute time has been reached. The H2O-3 JAR and Sparkling Water JAR files are obtained from H2O.ai and are specific to the Hadoop distribution. The administrator uploads the JAR files from the UI or performs a **secure copy** (**SCP**) to the Enterprise Steam filesystem.

- For specification details on installing Enterprise Steam, including requirements for Linux OS, Java, and Hadoop distributions, see the H2O Enterprise Steam documentation at `https://docs.h2o.ai/enterprise-steam/latest-stable/docs/install-docs/installation.html`.

Upgrading H2O-3 or Sparkling Water Is Easy (Again)

H2O-3 and Sparkling Water are not installed on the Hadoop cluster: here, they are wrapped into YARN jobs and distributed and run on the cluster as such. This makes upgrading easy: the administrator simply uploads a newer version of the JAR file to Enterprise Steam and the users launch their next H2O cluster with the newer versions. (Multiple versions can coexist, so users can select among them when launching an H2O cluster.)

For high availability on Hadoop, Enterprise Steam is installed as an active-passive setup with manual failover. This is described in the H2O documentation at `https://docs.h2o.ai/enterprise-steam/latest-stable/docs/install-docs/ha.html`.

Deployment as part of H2O AI Cloud

Enterprise Steam, H2O Core (H2O-3 and Sparkling Water) can be deployed as part of a larger H2O AI Cloud platform (*see Figure 12.2*). As such, it has a similar architecture to the two models just shown but is now integrated and installed as part of a larger H2O AI Cloud implementation. This will be discussed further in *Chapter 13, Introducing H2O AI Cloud.*

Now that we have explored H2O at Scale's enterprise architecture view, let's look at the security view.

H2O at Scale security

H2O at Scale has been deployed to numerous highly regulated enterprises in financial services, insurance, healthcare, and other industries. This typically involves extensive and thorough architecture and security reviews that vary across organizations. Key areas from an architecture standpoint are discussed in the following sections.

Data movement and privacy

Top security points for data movement and privacy are noted here:

- Data moves directly from the enterprise data source to the memory of the user's H2O cluster on the enterprise cluster and does not pass through H2O Enterprise Steam or the data scientist's client.

- Data in concurrent H2O clusters is isolated from other data: users cannot see or access each other's data during model building.

- Data in a user's H2O cluster is removed when the H2O cluster is stopped. Recall that data is partitioned in memory across a user's running H2O cluster nodes.

- Users can download data to an enterprise-storage-layer home directory (for example, in AWS S3 or HDFS) as configured by an administrator.

- Users cannot download data from an H2O cluster to a local disk.

User authentication and access control

The main points to note are outlined here:

- User authentication is done centrally through Enterprise Steam, which authenticates against the enterprise IdP. Enterprise Steam supports **OpenID**, **Lightweight Directory Access Protocol (LDAP)**, **Security Assertion Markup Language (SAML)**, and **Pluggable Authentication Modules (PAM)** authentication.

- User authentication through Enterprise Steam can be done through the H2O Steam **application programming interface (API)**. Users can generate **personal access tokens (PATs)** and use them as environment variables in their programs (as opposed to hardcoding them). The generation of a new PAT revokes the previous one. PATs are generated from the Enterprise Steam UI after the user logs in.

- Enterprise Steam supports running on a Kerberos-secured Hadoop cluster.

- For access to HDFS or Hive, Hadoop impersonation can be configured. This allows Enterprise Steam to appear to these data sources as the logged-in user and not as the Enterprise Steam service ID. This in turn allows H2O users to authenticate against these resources simply by logging in to Enterprise Steam (as opposed to explicitly passing authorization credentials).

- For JDBC and other access to data sources, the data scientist passes access credentials (for example, JDBC connection **Uniform Resource Locator** (**URL**), username, and password) as parameters to the H2O API's data import statement in the client IDE.

Enterprise Steam administrator configurations for user authentication and access control were covered in *Chapter 11, The Administrator and Operations View.*

Network and firewall

It is assumed that H2O at Scale is deployed securely behind the enterprise network either on-premises or in the cloud. Aspects of H2O network security are noted here:

- User access to Enterprise Steam and H2O clusters on the enterprise-server cluster is through **Hypertext Transfer Protocol Secure** (**HTTPS**). Enterprise Steam autogenerates a self-signed **Transport Layer Security** (**TLS**) certificate at installation time, but you will have the option to install your own after your first login. Additionally, certificates can be installed for LDAP, Kubernetes Ingress, and Kubernetes connection to MinIO and H2O.ai storage layers.

- Enterprise Steam exposes users to port 9555 by default, but this can be changed. Enterprise Steam runs behind a reverse proxy, and users never communicate directly with the H2O cluster itself.

Full details for H2O Enterprise Steam security can be found in the H2O documentation at `https://docs.h2o.ai/enterprise-steam/latest-stable/docs/install-docs/index.html`.

We have now explored H2O at Scale enterprise and security architecture from the perspective of their respective stakeholders. Let's now view how this relates to the data scientist.

The data scientist's view of enterprise and security architecture

Data scientists typically do not interact directly or frequently with enterprise and security architects. Nevertheless, data scientists may be impacted by these stakeholders in the way shown in the following diagram:

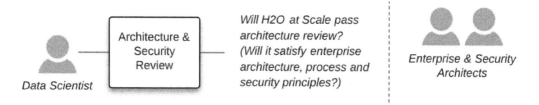

Figure 12.7 – Data-science view of enterprise and security architecture

Implementing H2O at Scale for the first time in the enterprise typically requires formal architecture and security review of the technology. Enterprise systems are large ecosystems of multiple technologies integrated into a coherent whole. Architect and security reviewers determine if H2O at Scale can be implemented in this ecosystem while conforming to principles that define the coherent whole. These principles are rules and guidelines defined by enterprise stakeholders that may range from guaranteeing new technology aligns with business needs, to guaranteeing the technology satisfies an extensive list of security requirements, to guaranteeing that the architecture enables **business continuity** (**BC**) after a disruption. These are just a few examples.

H2O at Scale technology has passed architecture and security reviews of *Fortune 100* and *Fortune 500* enterprises for reasons that have been outlined in this chapter. H2O at Scale can be summarized as follows:

- Has a light installation footprint and is easy to upgrade
- Integrates natively into existing frameworks (Kubernetes; YARN)
- Governs H2O users and H2O software management centrally through Enterprise Steam
- Provides high availability for BC
- Uses common communication protocols
- Implements extensive and well-known security mechanisms
- Ensures data privacy
- Provides support with known **service-level agreements** (**SLAs**)

Let's summarize what we learned in this chapter.

Summary

In this chapter, we examined the architecture and security underpinnings of H2O at Scale model-building technology. Such an examination provides architecture and security stakeholders a sound starting point to evaluate H2O against technical requirements and architecture principles that define enterprise systems. In general, H2O at Scale integrates and operates nicely in enterprise ecosystems and has a small software footprint in doing so. H2O also has extensive security capabilities that meet the high security demands of enterprises.

In the next chapter, we will take everything we have learned up to this point and place it in the larger context of H2O.ai's exciting new end-to-end machine learning platform called the H2O AI Cloud.

Section 5 – Broadening the View – Data to AI Applications with the H2O AI Cloud Platform

In this section, we will introduce the full-featured H2O AI Cloud platform. Importantly, we will recognize that what we have learned about in this book (ML at scale with H2O) can be implemented on its own or out of the box as a part of the larger H2O AI Cloud. We will first familiarize ourselves with the H2O AI Cloud components and where H2O-3, Sparkling Water, Enterprise Steam, and the MOJO fit. We will take this new knowledge and iterate examples of exciting new ways to build ML workflows and solutions using the H2O AI Cloud.

This section comprises the following chapters:

- *Chapter 13, Introducing H2O AI Cloud*
- *Chapter 14, H2O at Scale in a Larger Platform Context*

13
Introducing H2O AI Cloud

In the previous sections of this book, we explored in great detail how to build accurate and trustworthy **machine learning** (**ML**) models on massive data volumes using H2O technology, and how to deploy these models for scoring on a diversity of enterprise systems. In doing so, we became familiar with the technologies of H2O Core (H2O-3 and H2O Sparkling Water) and its distributed in-memory architecture to perform model building steps in a horizontally scalable way, using familiar IDEs and languages. We got to know H2O Enterprise Steam as a tool for data scientists to easily provision H2O environments and for administrators to manage users. We learned the technical nature of the H2O MOJO, the ready-to-deploy scoring artifact generated and exported from built models, and we learned a great diversity of patterns for scoring MOJOs on diverse target systems, whether real-time, batch, or streaming. We also learned how enterprise stakeholders beyond data scientists view and interact with H2O at scale technology.

In this chapter, we will expand our knowledge by learning that H2O offers a larger end-to-end ML platform called H2O AI Cloud that includes multiple specialized model building engines, an MLOps platform to deploy and monitor models, a feature store to share features for model building and scoring, and a technology layer often not considered in the context of ML platforms – a low-code SDK to easily build AI applications on top of rest of the platform and an App Store to host them.

Importantly, we will see that the technologies and skills we have learned up until now are actually a subset of the larger H2O AI Cloud.

In this chapter, we're going to cover the following main topics:

- An H2O AI Cloud overview
- An H2O AI Cloud component breakdown
- H2O AI Cloud architecture

Technical requirements

You can sign up for a 90-day trial to the H2O AI Cloud by visiting `https://h2o.ai/freetrial`. This will allow you to use the components of the platform with your own data or with trial data supplied by H2O.

We will see that part of the H2O AI Cloud is the ability of data scientists to build AI applications using an open source low-code SDK called H2O Wave. You can start building your own H2O Wave AI applications on your local machine by visiting here: `https://wave.h2o.ai/docs/installation`.

An H2O AI Cloud overview

The H2O AI Cloud is an end-to-end ML platform designed to enable teams to seamlessly work through building models, trusting models, and deploying, monitoring, and governing models. In addition, the H2O AI Cloud includes an AI application development and hosting layer to allow various personas to interact with all steps in an ML life cycle – from applications expressing sophisticated visualizations to user interactions and workflows. The application SDK allows data scientists and ML engineers (and traditional software developers) to quickly prototype, finalize, and publish AI applications in a purpose-built way. For example, applications can be built for business users to view dashboards of customer churn predictions with analytics on reason codes and then respond to high churn candidates. Data scientists, on the other hand, can use an AI application to interactively validate model predictions against subsequent ground truth and track analytics around that. Alternatively, data scientists and ML engineers can use an AI application to automate retraining pipelines by orchestrating data drift alerts with model retraining and redeployment while tracking analytics and auditing.

This simplified ML life cycle with an AI application layer is shown in the following diagram, and the H2O AI Cloud is organized around these layers:

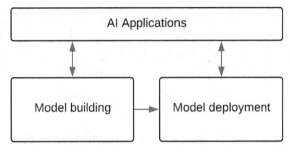

Figure 13.1 – A simplified ML life cycle with an AI app layer

H2O has built a modular, flexible, and fully capable end-to-end ML platform around this representation. The following diagram delineates the components of the H2O AI Cloud mapped to this life cycle:

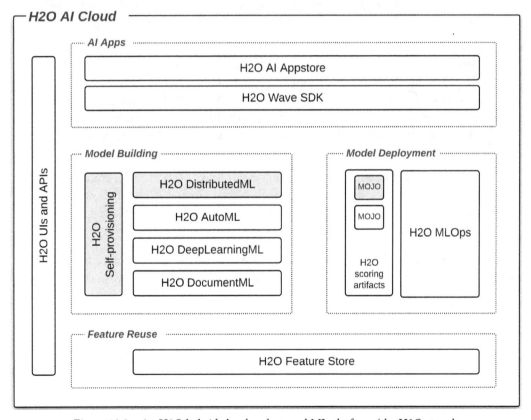

Figure 13.2 – An H2O hybrid cloud end-to-end ML platform (the H2O at scale components shown in gray)

Before diving into each component and its capabilities, let's first get a high-level understanding:

- **Model building**: There are four separate and specialized model building engines and a tool for data scientists to self-service provision their environments and for administrators to manage and govern users. Each model building engine generates a ready-to-deploy scoring artifact for models that are built.

- **Model deployment**: An MLOps component is used to deploy, monitor, manage, and govern models.

- **Feature store**: A feature store is available to reuse features both across teams during model building and across models during scoring.

- **AI applications**: A low-code SDK is available to rapidly build, prototype, and then publish AI applications. The SDK includes widgets and templates to build sophisticated and interactive visualizations and workflows. Data scientists and ML engineers build the application in a familiar code-based way, focusing mostly on organizing and feeding data to templates and widgets while ignoring the complexities of web applications.

- **AI App Store**: AI applications are developed locally and then published to an AI App Store component for consumption by business, data science, and other enterprise stakeholders. Clinicians in healthcare, for example, may use an application to prevent patients from being discharged from the hospital prematurely, while business analysts use a different part of the application to understand how frequent this is predicted to happen and why.

- **UI and API access to components**: Users can interact with H2O AI components interactively from both the UI and through APIs. Component APIs allow programmatic and automated approaches to interacting with the platform and stitching components together in unique ways.

In the next section, we will understand each H2O AI Cloud component more fully. Before doing so, however, let's introduce ourselves to the components with a table overview to get our bearings:

Component	Primary User(s)	Purpose
DistributedML *H2O-3,* *H2O Sparkling Water*	Data scientist junior to advanced	Model training on massive data volumes
AutoML *H2O Driverless AI*	Data scientist beginner to advanced	Highly automated fast time to accurate and trusted models
DeepLearningML *H2O Hydrogen Torch*	Data scientist beginner to advanced	Easy yet full-featured deep learning model building for multiple image and text problem types
DocumentML *H2O Document AI*	Data scientist, Document processing engineer	Document page classification and entity extraction where model understands document structure (for example, tables and page sections)
Self-provisioning Service *H2O Enterprise Steam*	Data scientist, Administrator	Self-provisioning of data science environments with administrator management and governance of users
Feature Store *H2O Feature Store*	Data scientist, ML operations	Store, update, share and operationalize features and their engineering for both model building and scoring
MLOPs *H2O MLOps*	ML operations	Model deployment, monitoring, management, and governance
Low Code SDK for AI Applications *H2O Wave*	Data scientist, ML engineer, Software developer	Low-code SDK to quickly build AI applications with sophisticated visualizations, user interactions and workflows while integrating H2O and non-H2O components
App Store *H2O AI Appstore*	AI application end users	Wave application hosting on an App Store for consumption across the organization

Figure 13.3 – A table summarizing H2O AI Cloud components

Finally, we need to relate H2O AI Cloud components to the focus of this book, which we will do in the following note.

> **How the Focus of This Book Relates to H2O AI Cloud**
>
> The focus of this book has been *ML at scale with H2O*, which alternatively has been called *H2O at scale*. We have focused on building ML models against massive datasets and deploying models to a diversity of enterprise scoring environments.
>
> From a component standpoint, the focus has been on H2O Core (H2O-3 and H2O Sparkling Water), H2O Enterprise Steam, and the H2O MOJO. These components can be deployed either as (a) separate from H2O AI Cloud, or (b) as members of H2O AI Cloud, as shown in *Figure 13.2*. See *Chapter 12, The Enterprise Architect and Security Views*, for an elaboration of this point.

Now that we understand the fundamentals of H2O AI Cloud, its components, and how they relate to the focus of this book, let's expand our view and ML capabilities by elaborating further on each component.

H2O AI Cloud component breakdown

Let's take a deeper dive into each of the components.

DistributedML (H2O-3 and H2O Sparkling Water)

DistributedML has been the focus of model building for this book, where it is called H2O Core to represent either H2O-3 or Sparkling Water in that context. Fundamentally, you use H2O Core to build models on massive datasets.

For the purposes of this chapter, the main features and capabilities are presented in the upcoming subsection For more details, see *Chapter 2, Platform Components and Key Concepts*, to review the distributed in-memory architecture that enables model building on a massive scale. See *Chapter 4, H2O Model Building at Scale – Capability Articulation*, to review its main capabilities in greater detail.

Key features and capabilities

The key features and capabilities of H2O Core (H2O-3 and Sparkling Water) are as follows:

- **Model building on massive data volumes**: H2O Core has an architecture that partitions and distributes data into memory across multiple servers. Model building computation is done in parallel against this architecture, thus achieving scaling needs for massive datasets. The larger the dataset, the more horizontally scaled the architecture will be.

- **Familiar data science experience**: Data scientists build H2O models using familiar IDEs and languages (for example, Python in Jupyter notebooks) to express the H2O model building API. The API hides the complexities of the H2O scalable architecture from the user. To a data scientist, the experience fundamentally is that of writing code against data frames.

- **Flexible data ingest**: H2O Core has connectors to access diverse data sources and data formats. Data is transferred directly from source to H2O Core-distributed memory.

- **Scalable data manipulation**: Data is manipulated in the distributed architecture and thus is done at scale. The H2O API makes data manipulation steps concise. Sparkling Water specifically allows data manipulation using Spark APIs (for example, Spark SQL) and the conversion of Spark DataFrames to H2OFrames in the same coding workflow.

- **State-of-the-art algorithms**: H2O Core implements state-of-the-art ML algorithms for supervised and unsupervised problems, including, for example, XGBoost, a **Gradient Boosting Machine (GBM)**, **Generalized Linear Model (GLM)**, and **Cox Proportional-Hazards (CoxPH)**, to name a few. These algorithms are run on the distributed architecture to scale to massive datasets.

- **AutoML**: H2O can build models using an AutoML framework that explores algorithm and hyperparameter space to build a leaderboard of best models. The AutoML framework is controllable through numerous settings.

- **Explainability and auto-documentation**: H2O Core implements extensive explainability capabilities and can generate auto-documentation to thoroughly describe model building and explain the resulting models.

- **MOJO**: Models built on H2O Core generate a ready-to-deploy and low-latency scoring artifact called a MOJO that can be flexibly deployed to diverse target environments. This was discussed in great detail in *Chapter 9, Production Scoring and the H2O MOJO.*

Let's move on to H2O AI Cloud's next model building engine.

H2O AutoML (H2O Driverless AI)

H2O Driverless AI is a highly automated AutoML tool built in part by Kaggle Grandmaster data scientists to incorporate data science best practices and AI heuristics to find highly accurate models in short amounts of time. Some of its key capabilities are rich explainability features, a genetic algorithm to iterate to the best model, and exhaustive feature engineering and selection to derive and use new features. Let's investigate these key features and capabilities.

Key features and capabilities

The H2O Driverless AI key features and capabilities are as follows:

- **Problem types**: H2O Driverless AI builds both *supervised* and *unsupervised* models:

  - **Supervised learning**: For supervised learning on tabular data, H2O Driverless AI addresses regression, binary and multiclass classification, and time-series forecasting problems. For supervised learning on images, Driverless AI addresses image classification, and for **natural language processing (NLP)**, it addresses text classification and context tagging problems.

  - **Unsupervised learning**: For unsupervised learning, Driverless AI tackles anomaly detection, clustering, and dimensionality reduction problems.

- **GPU support**: Driverless AI can leverage GPUs for image and NLP problems, which run TensorFlow and PyTorch algorithms.

- **Genetic algorithm**: Driverless AI uses a proprietary genetic algorithm to iterate across dozens of models, each of which varies in its algorithm (for example, XGBoost, Generalized Linear Model, and LightGBM), its exploration of hyperparameter space, and its exploration of feature engineering space. The best models are promoted to the next iteration and new model variations are introduced during each iteration. This continues until it cannot find a better model based on the settings that users make.

- **Feature engineering**: During the genetic algorithm, Driverless AI applies dozens of transformers in exhaustive ways to engineer new features from those in the original dataset and determine which ones to include in the final model. These transformers are categorized as follows:

  - **Numeric**: These are mathematical operations among two or more original features – for example, subtracting two features or clustering multiple features for the dataset and measuring the distance to a specific cluster for each observation.

  - **Categorical**: These are transformations of category labels to numbers – for example, taking the average or frequency of the target variable for each category and assigning it to the category represented for each observation.

  - **Time and date**: These are transformations of time and date fields to alternative time and date representations – for example, converting the date to the day of the week.

- **Time series**: These transformations derive new features useful for time-series problems – for example, using a lag time for a feature value.

- **Text**: These transformations convert strings to alternative representations – for example, using pre-trained **Bidirectional Encoder Representations from Transformers (BERT)** models to generate new language representations.

- **Bring your own recipes**: In addition to access to extensive expert settings, data scientists can control the automated ML process by importing their own code, which H2O calls recipes. These custom recipes can take the following form – *scorer* (your own performance metric used to optimize models in the genetic algorithm), *feature engineering* (your own engineered feature), or *algorithm* (your choice of ML algorithm to supplement familiar Driverless AI out-of-the-box algorithms).

- **Interpretability (Explainability)**: Users can interact with diverse and full-featured interpretability techniques to explain the resulting models. These techniques can be applied at the *global* (entire model) or *local* (individual record) levels. These techniques include *surrogate* and *actual model* techniques, including K-Lime and Shapley, Decision Tree, Disparate Impact Analysis, Sensitivity Analysis, and Partial Dependence Plots. There are also explainers for time-series and NLP problems specifically.

- **Auto-documentation**: Each final model generated by the genetic algorithm creates extensively standardized (typically over 60 pages) auto-documentation that describes in great detail experiment overview, data overview, methodology, validation strategy, model tuning, feature transformations and evolution, a final model, and explainability. The document is in paragraph, tabular, and graphic form.

- **MOJO**: Each final model generated by the genetic algorithm creates a ready-to-deploy and low-latency MOJO that is flexibly deployed to diverse target environments. This is a similar technology to that discussed in *Chapter 9, Production Scoring and the H2O MOJO*, for H2O at scale (H2O-3 and Sparkling Water).

> **Important Note**
> The MOJO for Driverless AI performs the feature engineering for features derived during the automated model building process.

Let's now move on to the DeepLearningML engine.

DeepLearningML (H2O Hydrogen Torch)

H2O Hydrogen Torch is a UI-based deep learning engine that empowers data scientists of all skill levels (and perhaps analysts for some use cases) to easily build state-of-the-art computer vision and NLP models. The key features and capabilities are as follows.

Key features and capabilities

The H2O Hydrogen Torch features and capabilities are as follows:

- **Problem types**: Currently, Hydrogen Torch addresses six **computer vision** (**CV**) and five NLP problem types, described briefly as follows:

  - **Image classification (CV)**: Images are classified into one or more sets of classes – for example, an image is classified as car versus truck.

  - **Image regression (CV)**: A continuous value is predicted from an image – for example, the steering angle from a self-driving car image is positive 20 degrees from the center line.

  - **Object detection (CV)**: An object (or objects) is classified from an image and its position coordinates are identified as a bounding box – for example, multiple cars are identified, each with a rectangle defined around it.

  - **Semantic segmentation (CV)**: An object (or objects) is classified as well as its exact shape, defined by pixel positions – for example, the exact outline of a person or all people in an image.

  - **Instance segmentation (CV)**: This is the same as semantic segmentation, but when multiple objects of the same class are identified in instance segmentation, they are treated separately, whereas in semantic segmentation, they are treated as one object.

  - **Image metric learning (CV)**: Predicts the similarity between images – for example, for a picture of a retail product, it will find the likelihood that a new picture is the same product.

  - **Text classification (NLP)**: Classifies text (document, page, and snippet) into a class – for example, classifying the sentiment or intent of text.

  - **Text regression (NLP)**: Predicts a continuous value from text – for example, prediction of a person's salary from a resume.

  - **Text sequence to sequence (NLP)**: Converts text sequences in one context to text sequences in another context – for example, converting a document into a summary.

- **Text token classification (NLP)**: Classifies each word in a text to a label – for example, identifying the United Nations as an organization (an example of **Named Entity Recognition** (**NER**)) or identifying a word as a noun or verb (example of **Part-of-Speech** (**POS**) tagging).

- **Text metric learning (NLP)**: Predicts the similarity between two sets of text – for example, identifying duplicate information or similar documents.

- **Ease of building deep learning models**: Hydrogen Torch is a no-code approach to building deep learning models. The user interacts with a UI that has extensive controls on hyperparameter tuning and a rich interface to quickly iterate, understand, and evaluate model outcomes. Models can be exported for deployment to Python or H2O MLOps environments.

- **Modes for user skill set**: The Hydrogen Torch training UI adapts to the user skill level by exposing fewer or more model building settings, according to whether the user is a novice, skilled, an expert, or a master.

Now, let's move on to a model building engine that focuses on documents.

DocumentML (H2O Document AI)

Documents typically represent a vast untapped data source for enterprises to apply ML techniques to automate processing steps, and thus save large amounts of time and money compared to manual processing. H2O's Document AI engine learns from documents to accomplish this automation.

Document AI goes beyond simple **Optical Character Recognition** (**OCR**) and NLP by learning to recognize information structures of documents such as tables, forms, logos, and sections. The Document AI model is trained to extract text entities from documents using these capabilities. Documents can thus be processed to extract specific information from medical lab results, financial statements, loan applications, and so on. This output can then drive analytics and workflows from these documents, which become increasingly more valuable as the volume of document processing grows. Document AI can also classify an entire document to further automation of document processing pipelines.

Key features and capabilities

Let's breakdown these capabilities further:

- **Document ingest**: Ingests documents such as PDFs, images, Word, HTML, CSV files, text files, emails, and others.

- **Preprocessing**: Document AI uses OCR and NLP capabilities to perform multiple preprocessing steps, such as handling embedded text (for example, PDF metadata) and logos, and orientating, deskewing, and cropping pages.

- **Apply document labels**: Users access a UI to apply labels to document text. Models will be trained to recognize these labeled entities. For example, on a medical lab document, the user applies labels to the patient name, the lab name, the lab address, the test name, the test result value, the test result unit, the test result normal range, and so on.

- **Train models**: Document AI trains against a labeled document set. It learns to associate text with labels in the larger context of the structure of the document – for example, lab results are reported from rows in a table. Note that models are trained against a known document set and afterward will be able to pull information from documents they have never seen before. For example, each lab produces its own report (its own design, the styling of tables, number of pages, the position of the patient name in the document, and so on). Even though the model is trained on a small set of lab reports (typically 100 or so), it can then pull information from documents sent from a lab it has not been trained on.

- **Post-processing**: Document AI allows users to customize and standardize how results are outputted. For example, users can define an output JSON structure with date output formats standardized.

- **Model deployment**: Models can be exported and deployed to H2O MLOps or a Python environment of your choice.

Now that we have explored the four specialized model building engines on H2O AI Cloud, let's see how features for those engines can be shared and operationalized.

A self-provisioning service (H2O Enterprise Steam)

H2O Enterprise Steam allows users to self-provision model building environments and administrators to govern users and their resource consumption. As with H2O Core and the scoring artifact it generates called the MOJO, Enterprise Steam is considered a key component of H2O ML at scale and was introduced in *Chapter 2, Platform Components and Key Concepts*, and then explored in detail in *Chapter 11, The Administrator and Operations View*.

Note that in that context, Enterprise Steam was used to self-provision and manage H2O Core environments only, but in the context of the H2O AI Cloud, it is used to manage all H2O model building engines. Let's review its key capabilities.

Key features and capabilities

The key capabilities of H2O Enterprise Steam are listed briefly as follows:

- **Easy self-provisioning of H2O model building environments**: Data scientists can define, launch, and manage their H2O model building environments from the Enterprise Steam UI or API. Note that, currently, this is true for DistributedML (H2O Core) and AutoML (Driverless AI) environments. Hydrogen Torch and Document AI environments currently are launched as applications, but they are road-mapped to consolidate into the Enterprise Steam self-provisioning framework.

- **Administrator management and governance of users**: Administrators manage users and define the amount of resources (CPU and memory) they can use when provisioning environments, including how long those environments sit idle before spinning down.

Let's move on now to the Feature Store component.

Feature Store (H2O AI Feature Store)

H2O AI Feature Store is a system to organize, govern, share, and operationalize predictive ML features across the enterprise in both the model building and live scoring contexts. This saves significant time for data scientists to discover features and for both data scientists and ML engineers to transform raw data into these features. Let's explore the capabilities further.

Key features and capabilities

Here are some key features of the H2O AI Feature Store:

- **Versatile feature publishing and search workflow**: Data scientists and engineers engineer feature pipelines using pre-built integrations into Snowflake, Databricks, H2O Sparkling Water, and other technologies. The resulting features are outputted to the H2O AI Feature Store with over 40 metadata attributes associated with the feature. This cataloging of features and their attributes allows other data scientists to search for relevant features and for the Feature Store's built-in AI to recommend features.

- **Scalable and timely feature consumption**: Each feature in the Feature Store has a defined duration until it is refreshed. Features can be stored offline for training and batch scoring or stored online for low-latency real-time scoring.

- **Automatic feature drift and bias detection**: Features are automatically checked for data drift and users are alerted when drift is detected. This can be essential in deciding to retrain models with more recent data. Features are also automatically checked for bias and alert users when bias is detected. This can be essential in retraining models to remove bias.

- **Access management and governance**: H2O AI Feature Store integrates with the enterprise identity provider to authenticate users and authorize access to features. Features and their metadata are versioned for regulatory compliance and to backtest models against ground truth.

H2O AI Cloud has a fully capable model operations component. Let's learn more about that next.

MLOps (H2O MLOps)

H2O MLOps is a platform to deploy, manage, monitor, and govern models. These can be either models generated from any of the H2O model building engines (DistributedML, AutoML, DeepLearningML, or DocumentML) or models from non-H2O software (for example, scikit-learn or MLflow). Note that H2O MLOps workflows can be completed using the UI or API, with the latter essential for integrating into **continuous integration and continuous deployment (CI/CD)** workflows. Major capabilities are elaborated as follows.

Key features and capabilities

Here are the key features of H2O MLOps:

- **Model deployment**: Easy deployment of H2O and non-H2O models. Scoring is available as a REST endpoint for both real-time and batch scoring. Models are deployed as either a single model (simple deployment), champion/challenger (compare a new model to current model where the only current model is live), or an A/B test (multiple live models with live data are routed among them in configured proportions). Models are deployed to defined environments, typically development and production, but you may add more.

- **Model monitoring**: Models are monitored for health, scoring latency, data drift, fairness (bias) degradation, and performance degradation. Alerts are presented on the monitoring dashboard and sent to configured recipients. Alerts can be used to trigger model retraining and deployment.

- **Model management**: Models can be compared and evaluated, promoted to a registry, and then deployed. Models are associated with extensive metadata, allowing traceability to model building details and evaluation against other models. Models in the registry (and subsequent deployment) are versioned. Deployed models can be rolled back to previous versions.

- **Model governance**: The versioning and traceability achieved through model management create a lineage of model history. Users have role-based access with actions that are audited. Administrators have a dedicated dashboard to provide visibility across all users, models, and audit logs. These capabilities combine a result in an overall governance process that minimizes model risk and facilitates regulatory compliance.

We started this chapter by recognizing an application layer that integrates the rest of the H2O AI Cloud platform. Let's learn more about that.

Low-code SDK for AI applications (H2O Wave)

H2O Wave is an open source and low-code Python SDK to build real-time AI applications with sophisticated visualizations. Low code is achieved by abstracting the complexities of web application coding away from the application developer while exposing higher-level UI components as templates, themes, and widgets. Data scientists and ML engineers are intended as developers (as well as software developers themselves).

Examples of H2O Wave applications have been built by H2O data scientists as capability demonstrators. These can be found on the H2O AI Cloud 90-day evaluation site at `https://h2o.ai/freetrial`. Additional examples are on the H2O public GitHub repository at `https://github.com/h2oai/wave-apps`.

How Do I Try Building Wave Applications?

Instructions to download the Wave server and SDK to build your own applications can be found at `https://wave.h2o.ai/docs/installation`.

Key features and capabilities

The following are the key features and capabilities of H2O Wave:

- **Low-code SDK**: Data scientists and ML engineers focus on specifying templates and widgets and feeding data into them to create sophisticated visualizations, dashboards, and workflows. The complexities of web application code are abstracted away from the developer.

- **Extensive native data connectors**: You have access to over 160 connectors to data sources and sinks from the SDK.

- **Native H2O APIs**: The SDK includes H2O APIs that integrate other H2O AI Cloud components. This enables data scientists and ML engineers to integrate aspects of the ML life cycle as a backend to the application visualizations and workflows.

- **Use any Python package**: Applications are isolated as containers, thus allowing any Python package to be used by the application – for example, NumPy and pandas for data manipulation and Bokeh and Matplotlib for data visualizations, to name just a few.

- **Integrate non-H2O technology**: When Python packages in your application represent public APIs such as the Twitter API, AWS service APIs, or your own private Python APIs, Wave applications can integrate non-H2O technology into its visualizations and workflows. Wave applications can thus be built as single panes of glass across multiple technologies.

- **Publish to H2O App Store**: Wave applications are developed locally and then published to the H2O AI Cloud App Store for enterprise consumption.

Let's now take a look at the H2O App Store.

App Store (H2O AI App Store)

The H2O AI App Store hosts your H2O Wave applications in your H2O AI Cloud instance. H2O Wave applications are hosted in a searchable and role-based way. Users logged in to the App Store see only the applications they are allowed to use and can find them by custom-defined categories or by search. Wave application developers publish to the App Store, and administrators manage the App Store.

Application consumers thus access and use Wave applications through the App Store, though data scientists and ML engineer developers may prototype with consumers locally before publishing to the App Store.

Let's now get a high-level understanding of the H2O AI Cloud architecture.

H2O AI Cloud architecture

We will not dive deep into H2O AI Cloud Architecture but will review three important architecture points:

- **Components are modular and open**: The platform's modular architecture allows enterprises or groups to use the components they need and to hide and ignore the ones they do not. H2O AI Cloud is also open – its components can coexist and interact with the larger enterprise ecosystem, including non-H2O AI/ML components. The MLOps component, for example, can host non-H2O models, such as scikit-learn models, and the AI application Wave SDK can integrate non-H2O APIs with its own.

- **Cloud-native architecture**: H2O AI Cloud is built on a modern Kubernetes architecture that achieves efficient resource consumption among cloud servers. In addition, H2O workloads on the AI Cloud are ephemeral – they spin up when needed, spin down when not in use, and retain state when spinning up again. The H2O AI Cloud also leverages the cloud service providers' managed services – for example, using the cloud-managed Kubernetes service and maintaining state in a managed PostgreSQL database.

- **Flexible deployment**: H2O AI Cloud can be deployed in an enterprise's cloud, on-premises, or in a hybrid environment. Alternatively, it can be consumed as a managed service where H2O hosts and manages the enterprise's H2O AI Cloud platform in H2O's cloud environment.

These architecture points combined with the capabilities of each component mean that enterprises can fit the H2O AI Cloud to their specific environment, use case needs, and stage of their AI transformation journey.

Let's summarize what we've learned in this chapter.

Summary

In this chapter, we expanded our view beyond *H2O ML at scale*, which has been the focus of this book to this point. We did this by introducing H2O's end-to-end ML platform called H2O AI Cloud. This platform has a broad set of components in the model building and model deployment steps of the ML life cycle and introduces a lesser-considered layer to this flow – easy-to-build AI applications and an App Store to serve them. We learned that H2O AI Cloud has four specialized engines for building ML models – DistributedML, AutoML, DeepLearningML, and DocumentML. We learned that MLOps has a full capability set around deploying, monitoring, managing, and governing models for scoring. We also learned that a Feature Store is available to centralize and reuse features for model building and model scoring.

Importantly, we learned that the focus of this book, building ML models on massive datasets and deploying to enterprise systems for scoring (what we have called H2O at scale), uses technology (H2O Core, H2O Enterprise Steam, and H2O MOJO) that is actually a subset of the larger H2O AI Cloud platform.

We made the point thatH2O at scale technology can be deployed separately from H2O AI Cloud or as a part of the larger platform. In the next chapter, we are going to see additional capabilities that H2O at scale takes on by being a member of the H2O AI Cloud.

14
H2O at Scale in a Larger Platform Context

In the previous chapter, we broadened our view of H2O **machine learning** (**ML**) technology by introducing H2O AI Cloud, an **end-to-end** ML platform composed of multiple model-building engines, an MLOps platform for model deployment, monitoring, and management, a Feature Store for reusing and operationalizing model features, and a low-code **software development kit** (**SDK**) for building **artificial intelligence** (**AI**) applications on top of these components and hosting them on an app store for enterprise consumption. The focus of this book has been what we have called **H2O at scale**, or the use of H2O Core (H2O-3 and Sparkling Water) to build accurate and trusted models on massive datasets, H2O Enterprise Steam to manage H2O Core users and their environments, and the H2O MOJO to easily and flexibly deploy models to diverse target environments. We learned that these H2O-at-scale components are natively a part of the larger H2O AI Cloud platform, though they can be deployed separately from it.

In this chapter, we will explore how H2O at scale achieves greater capabilities as a member of the H2O AI Cloud platform. More specifically, we will cover the following topics:

- A quick recap of H2O AI Cloud
- Exploring a baseline reference solution for H2O at scale
- Exploring new possibilities for H2O at scale
- A Reference H2O Wave app as an enterprise AI integration fabric

Technical requirements

This link will get you started on developing H2O Wave applications: `https://wave.h2o.ai/docs/installation`. H2O Wave is open source and can be developed on your local machine. To get full familiarity with the H2O AI Cloud platform, you can sign up for a 90-day trial of H2O AI Cloud at `https://h2o.ai/freetrial`.

A quick recap of H2O AI Cloud

The goal of this chapter is to explore how H2O at scale, the focus of this book, picks up new capabilities when used as part of the H2O AI Cloud platform. Let's first have a quick review of H2O AI Cloud by revisiting the following diagram, which we encountered in the previous chapter:

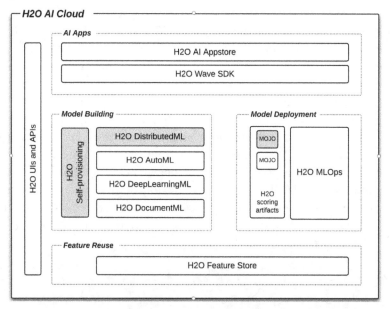

Figure 14.1 – Components of the H2O AI Cloud platform

As a quick summary, we see that H2O AI Cloud has four specialized model-building engines. H2O Core (H2O-3, H2O Sparkling Water) represents H2O DistributedML for horizontally scaling model building on massive datasets. H2O Enterprise Steam, in this context, represents a more generalized tool to manage and provision the model-building engines.

We see that the H2O MOJO, exported from H2O Core model building, can be deployed directly to the H2O MLOps model deployment, monitoring, and management platform (though, as seen in *Chapter 10, H2O Model Deployment Patterns*, the MOJO can be deployed openly to other targets as well). Note that the H2O specialized **automated ML** (**AutoML**) engine called Driverless AI also produces a MOJO and can be deployed in ways shown in *Chapter 10, H2O Model Deployment Patterns*. We also see that the H2O AI Feature Store is available to share and operationalize features both in the model-building and model-deployment contexts.

Finally, we see that H2O Wave SDK is available to build AI apps over the other components of the H2O AI Cloud platform and then publish to an H2O App Store as part of the platform.

Let's now start to put these pieces together into various H2O-at-scale solutions.

Exploring a baseline reference solution for H2O at scale

So, let's now explore how H2O-at-scale components benefit from participating in the H2O AI Cloud platform. To do so, let's first start with a baseline solution of H2O at scale outside of H2O AI Cloud. The baseline solution is shown in the following diagram. We will use this baseline to compare solutions where H2O at scale does integrate with AI Cloud components:

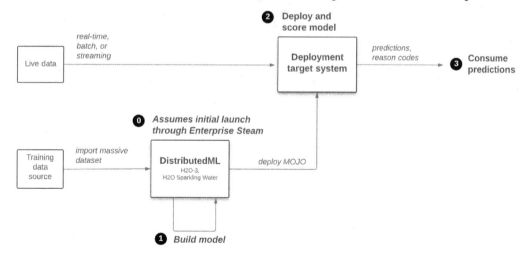

Figure 14.2 – Baseline solution for H2O at scale

> **Important Note**
>
> For this and all solutions in the chapter, it is assumed that the data scientist
> used H2O Enterprise Steam to launch an H2O-3 or H2O Sparkling Water
> environment. See *Chapter 3, Fundamental Workflow – Data to Deployable
> Model*, for an overview of this step.

A quick walkthrough of its solution flow is summarized as follows:

1. The data scientist imports a large dataset and uses it to build an ML model at scale.
   See the chapters in *Part 2, Building State-of-the-Art Models on Large Data Volumes
   Using H2O*, for a deep exploration of this topic.

2. The operations group deploys the model artifact (called the H2O MOJO) to a
   scoring environment. See *Chapter 10, H2O Model Deployment Patterns*, to explore a
   diversity of such target systems for H2O model deployment.

3. The predictions are consumed and acted up by software or tooling within a business
   context. (Here, we are assuming the deployed model is a **supervised learning**
   model, though it could be an unsupervised model that generated outputs that are
   not predictions; for example, cluster membership of an input).

Let's use this baseline solution to start adding H2O AI Cloud components and thus
see how H2O at scale gains additional capabilities through membership of this larger
platform.

Exploring new possibilities for H2O at scale

Now, let's step through different ways we can integrate H2O at scale—the focus of
this book—with the rest of the H2O AI Cloud platform and thereby achieve greater
capabilities and value.

Leveraging H2O Driverless AI for prototyping and feature discovery

H2O's AutoML Driverless AI component is a highly automated model-building tool that
uses (among other features) a genetic algorithm, AI heuristics, and exhaustive automated
feature engineering to build accurate and explainable models— typically in hours—that
are then deployed to production systems. Driverless AI, however, does not scale to train
on the hundreds of GB to TBs sized datasets that H2O-3 and Sparkling Water handle.

It is quite useful, however, for data scientists to feed sampled data from these massive datasets to Driverless AI and then use the AutoML tool to (a) quickly prototype the model to gain an early understanding and (b) discover auto-engineered features that contribute to an accurate model but would otherwise be difficult to find from pure manual and domain knowledge means. The resulting knowledge from Driverless AI in this workflow is then used as a starting point to build models at scale against the original unsampled data using H2O-3 or Sparkling Water.

This is shown in the following diagram and then summarized:

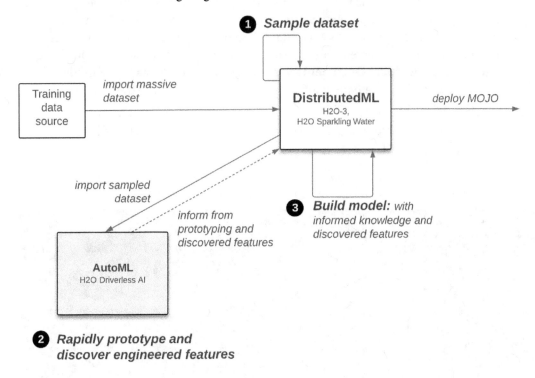

Figure 14.3 – Leveraging H2O Driverless AI for prototyping and feature discovery

The workflow steps to leverage Driverless AI to quickly prototype and discover features for model building at scale are provided here:

1. Import a large-volume dataset into the distributed in-memory architecture inherent in the H2O-at-scale environment. Sample the imported data into a smaller subset (typically 10 to 100 GB) using the H2O-3 `split_frame` method (here in Python) with an appropriate ratio defined as an input parameter to achieve the desired sample size. Write the output to a staging location that Driverless AI can access.

2. Import the sampled dataset to Driverless AI. Use defaults to quickly prototype an accurate model. Use different settings to continue prototyping based on your domain and data-science experience. Explore explainability techniques on models. Explore engineered features with the highest contribution to models. Use **Automated Model Documentation** (**AutoDoc**) to understand the models more deeply and translate the names of engineered features into their underlying mathematical and logical representations.

3. Use the knowledge from *Step 2* to guide your model building against the full massive dataset on H2O-3 or Sparkling Water.

Driverless AI was overviewed in *Chapter 13, Introducing H2O AI Cloud*. The following screenshot shows an experiment iterating toward a final accurate model:

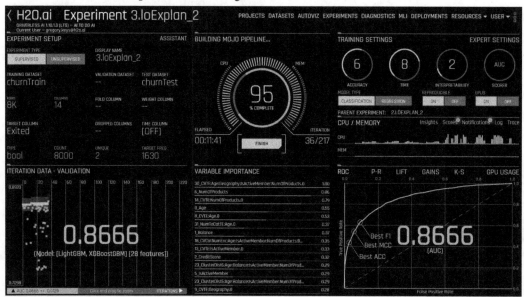

Figure 14.4 – Driverless AI finding an accurate model

Note the lower-left panel of *Figure 14.4*, which shows the progress of the genetic algorithm iterating across models. Each square is a separate ML model that has been built. Each of these models uses one of an automated choice of algorithms (for example, XGBoost; **Light Gradient Boosting Model (LightGBM)**; **Generalized Linear Model (GLM)**) that explores an extremely wide hyperparameter space (for example, combinations of learning rates, tree depths, number of trees, to name a few).

Importantly, each model built by the genetic algorithm also explores an extremely wide space of features engineered from those in the original imported dataset. The experiment will stop with a final best model, typically a stacked ensemble of preceding top models. Users can run many experiments in short amounts of time with the intent of exploring, by changing many high-level and low-level settings and evaluating outcomes.

Knowledge gained from these rapid explorations, including feature engineering that is important to the final model, can be used as a starting point to build models at scale with H2O-3 or Sparkling Water.

Note that for smaller datasets, Driverless AI is quite effective at finding highly predictive models in short periods of time. H2O-3 and Sparkling Water, however, are needed for scaling to massive datasets or for taking a more controlled code-based approach to model building. As shown here, for the code-based approach, it is valuable to first prototype a problem with Driverless AI and then use the resulting insights and engineered features as a guide to the code-based approach.

Driverless AI (AutoML) versus H2O-3 (DistributedML): When to Use Which?

Driverless AI is a highly automated **user interface** (**UI**)-based (or **application programming interface** (**API**)-based) AutoML component of H2O AI Cloud designed to quickly find accurate and trusted models for production scoring. Use it when you want a highly automated approach (with extensive user controls) to model building and when dataset sizes are less than 100 GB (though more resource-heavy server instances can work with larger datasets).

Use H2O-3 or Sparkling Water when your datasets are greater than 100 GB (and into TBs) or for a code-based approach to model building when you want more control of the model-building process. Note that H2O-3 and Sparkling Water have AutoML capabilities, as described in *Chapter 5, Advanced Model Building – Part 1*, but those in Driverless AI are far more sophisticated, extensive, and automated.

Use Driverless AI to prototype a problem and take the resulting insights and discovery of engineered features to guide your model building on H2O-3 or Sparkling Water.

We have seen our first example of H2O-at-scale model building gaining capabilities by interacting with a component in the H2O AI Cloud platform. Let's now look at our next example.

Integrating H2O MLOps for model monitoring, management, and governance

Models built with H2O-3 and Sparkling Water generate their own ready-to-deploy low-latency scoring artifact called the H2O MOJO. As we showed in *Chapter 10, H2O Model Deployment Patterns*, this scoring artifact can be deployed to a great diversity of production systems, ranging from real-time scoring from **REpresentational State Transfer (REST)** servers and batch scoring from databases to scoring from streaming queues (to name a few). We also showed in that chapter that deploying to a REST server provides a useful integration pattern for integrating scored predictions into common **business intelligence (BI)** tools such as Microsoft Excel or Tableau.

The H2O MLOps component of H2O AI Cloud is an excellent choice for deploying H2O-3 or Sparkling Water models (or Driverless AI models and non-H2O models—for example, scikit-learn models, for that matter). The integration is quite simple, as shown in the following diagram:

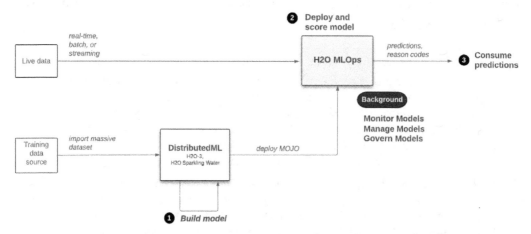

Figure 14.5 – Deployment of H2O-at-scale models to the H2O MLOps platform

The steps to integrate these models are straightforward, as outlined here:

1. Build your model with H2O-3 or Sparkling Water and export the MOJO for deployment.

2. From H2O MLOps, deploy the staged MOJO either from the UI or the MLOps API. Recall from *Chapter 13*, *Introducing H2O AI Cloud*, that there are many deployment options, including real-time versus batch, single model versus champion/challenger, or A/B testing.

3. The model is now scorable from a unique REST endpoint. Predictions with optional reason codes are ready to be consumed by your system, whether that is a web application, a BI tool, and so on.

For all models during and after this flow, MLOps performs important tasks around monitoring, managing, and governing models. The following screenshot, for example, shows the data drift monitoring screen for H2O MLOps:

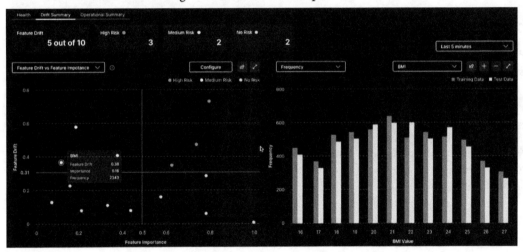

Figure 14.6 – Model-monitoring screen for H2O MLOps

The purpose of monitoring data drift is to detect whether the distribution of feature values in live scoring data is diverging from that in the training data from which the model was built. The presence of data drift suggests or indicates that the model should be retrained with more recent data and then redeployed to align with current scoring data.

In the lower-left panel in *Figure 14.6*, data drift for all model features is represented in two dimensions: drift on the vertical axis and feature importance on the horizontal axis. This view allows partitioning of drift into quadrants of drift importance, with high drift and high feature importance being the most important drift. Multiple settings are available to define the drift statistic, the time frame of measurement, and other aspects of viewing drift. There is also a workflow to configure automated alert messaging for drift. These can be used either for data scientists to manually decide on whether to retrain a model or for fully automated model retraining and deployment through H2O APIs.

Drift detection is just one of the capabilities of H2O MLOps. See the H2O documentation at `https://docs.h2o.ai/mlops/` for a full description of H2O MLOps model deployment, monitoring, management, and governance capabilities.

Let's now look at how H2O-3 and Sparkling Water can integrate into H2O AI Feature Store.

Leveraging H2O AI Feature Store for feature operationalization and reuse

Enterprises often achieve economy of scale by centralizing and sharing environments or assets across the organization. H2O AI Feature Store achieves economy of scale by centralizing model features and the operationalization of their curation through engineering pipelines for reuse across the organization.

Reuse through the Feature Store occurs during both model-building and model-scoring contexts. For example, let's say that a valuable feature across the organization is the percentage change in asset price compared to the previous day. Imagine, though, that asset price is stored as price per day and asset prices are stored in multiple source systems. Feature Store handles retrieval of features from source systems and calculation of new values from the original (that is, the feature engineering pipeline) and caching the result (the engineered feature) to be shared for model training and model scoring. Model building uses an offline mode—that is, batched historical data—and uses an online mode for model scoring—that is, recent data. This is shown in the following diagram:

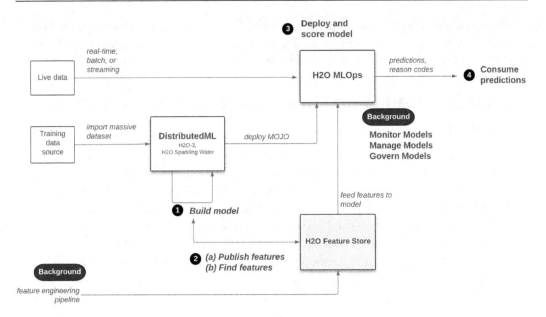

Figure 14.7 – Integration of H2O AI Feature Store with H2O-3 or Sparkling Water

Here's a summary of the workflow in *Figure 14.7*:

1. A data scientist builds a model.

2. During the model-building stage, the data scientist may interact with the feature store in two different ways, as outlined here:

    ▪ The data scientist may publish a feature to the Feature Store and include associated metadata to assist in search and operationalization by others. An ML engineer operationalizes the engineering of the feature from the data source(s) to feed the feature values in the feature store. Each feature is configured for how long it lives before being updated—for example, update each minute, day, or month.

    ▪ The data scientist may search the Feature Store for features during model building. They use the Feature Store API to import the feature and its value into the training data loaded by H2O-3 or Sparkling Water (more specifically, into the in-memory H2OFrame, which is analogous to a DataFrame).

3. The model is deployed to H2O MLOps, which is configured to consume the feature from the feature store. The feature is updated at its configured interval.

4. The predictions and—optionally—reason codes are consumed.

We have ended our workflows with predictions being consumed. Let's now showcase how the H2O Wave SDK can be used by data scientists and ML engineers to quickly build AI applications with sophisticated visualizations and workflows around model predictions.

Consuming predictions in a business context from a Wave AI app

ML models ultimately gain value when their outputs are consumed by personas or automation executing workflows. These workflows can be based on single predictions and underlying reason codes themselves or from insights and intelligence gained from them. For example, a customer service representative identifies customers who have a high likelihood of leaving the business and proactively reaches out to them with improvements or incentives to stay based on the reasons why the model made its prediction of likely churn. Alternatively, an analyst explores multiple interactive dashboards of churn predictions made in the past 6 months to gain insights into the causes of churn and identify where the business can improve to prevent churn.

As we learned in *Chapter 13*, *Introducing H2O AI Cloud*, H2O Wave is a low-code SDK used by data scientists and ML engineers to easily build AI applications and publish them to an App Store for enterprise or external use. What do we mean by an AI app? An AI app here is a web application that presents one or more stages of the ML life cycle as rich visualizations, user interactions, and workflow sequences. The H2O Wave SDK makes these easy to build by exposing UI elements (dashboard templates, dialogs, and widgets) as attributes-based Python code while abstracting away the complexities of building a web application.

In our example here, Wave apps are being used specifically by business personas to consume predictions, as shown in the following diagram. Note that our next example will be a Wave app used by data scientists to manage model retraining, and not to consume predictions:

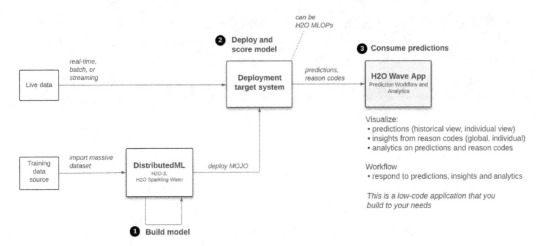

Figure 14.8 – Prediction consumption in Wave AI applications

The ML workflow in *Figure 14.8* is a familiar one but with a Wave app that consumes predictions. Wave's low-code SDK enables multiple integration protocols and thus allows predictions and reason codes to be consumed in real time from a REST endpoint or as a batch from a file upload or data warehouse connection, for example. As shown in the preceding diagram, from a business context, Wave apps that consume predictions can do the following:

- Visualize predictions from historical and individual prediction views
- Visualize insights from underlying prediction reason codes, from both global (model-level) and individual (single model-scoring) views
- Perform BI analytics on predictions and insights
- Perform workflows for humans in the loop to act on visualizations and analytics

Let's look at a specific example. The screenshot that follows shows one page of a Wave app that consumes and displays visualizations from predictions on employee churn—in other words, the predicted likelihood that employees will leave the company:

Figure 14.9 – A Wave app visualizing employee churn

This page shows views of the latest batch of employee-churn predictions. The upper-left panel shows the probability distribution of these predictions. We can see that most employees that were scored by the model have a low probability of leaving, though there is a long tail of higher-probability employees. The upper-right panel provides insights into why employees depart the company: it shows the Shapley values or feature contributions (reason codes) to the prediction. We see that overtime, job role, and shift schedule are the top three factors contributing to the model's predictions of churn. The bottom panels show visualizations to help understand these predictions better: the left panel displays a geographic distribution of churn likelihood, and the right shows a breakdown by job role. The thin panel above the map allows user interaction whereby the slider defines the probability threshold of classifying an employee as churn or no churn: the bottom two panels refresh accordingly as a result.

The previous screenshot of the Wave app visualized predictions at a batch level—that is, of all employees scored over a time period. The application also has a page that displays predictions visualized at the individual level. When the user clicks on an individual employee (displayed with associated churn probability and other employee data), Shapley values are displayed, showing the top features contributing to the likelihood of churn for *that individual*. A particular individual may show, for example, that monthly income is the larger contribution to the prediction and that overtime actually contributes to the individual *not* churning. This insight suggests that the employee may leave the company because they are not making enough money and are trying to make more. This allows the employee's manager to evaluate a salary increase to help guarantee they stay with the organization.

The UI in *Figure 14.9* shows predictions placed in a business context where individuals can act upon them. Keep in mind that H2O Wave is quite extensible and can incorporate Python packages of your liking, including Python APIs, to non-H2O components. Also, remember that the example Wave app shown here is meant to be a capability demonstrator: it is not an out-of-the-box point solution to manage employee churn but rather an example of how data scientists and ML engineers can easily build AI applications using the Wave SDK.

H2O Wave SDK Is Very Extensible

The UI in *Figure 14.9* is fairly simple but nevertheless effective in putting predictions into a business and analytical context. The H2O Wave SDK is quite extensible and thus allows greater layers of sophistication to be included in applications built from it.

You can, for example, implement HTML **Cascading Style Sheets** (**CSS**) to give the **user experience** (**UX**) a more modern or company-specific look. Because Wave applications are containerized and isolated from each other, you can install any Python package and use it in the application. You can, for example, implement **Bokeh** for powerful interactive visualizations or **pandas** for data manipulation, or a vendor or home-grown Python API to interact with parts of your technology ecosystem.

Note that the main intent of H2O Wave is for you to build your own AI applications with its SDK and to make them purpose-built for your needs. Applications are developed locally and can be prototyped quickly with intended users, then finalized, polished, and published to H2O App Store for enterprise role-based consumption. H2O will, however, provide example applications to you as code accelerators.

You can explore live Wave apps by signing up for a 90-day free trial of H2O AI Cloud at `https://h2o.ai/freetrial/`. You can also explore and use the H2O Wave SDK, which is open source, by visiting `https://wave.h2o.ai/`.

We have just explored how Wave apps can be built to consume predictions as part of a business analytic and decision-making workflow. Business users need not be the only Wave app users. Let's now look at a Wave app used by data scientists and built to drive the model-building and model-deployment stages of the ML life cycle.

Integrating an automated retraining pipeline in a Wave AI app

The H2O Wave SDK includes native APIs to other H2O AI Cloud components. This allows data scientists to build Wave apps to accomplish data-science workflows (compared to building applications in a business-user context, as shown in the previous example).

A common need in data science, for example, is to recognize data drift in deployed models and then retrain the model with recent data and redeploy the updated model. The following diagram shows how this can be done using a Wave app as both an automation orchestrator and UI for tracking the history of retraining and performing analytics around the history. This is an application idea and can be defined differently, but the general idea should be helpful.

Here, you can see an overview of the full ML workflow:

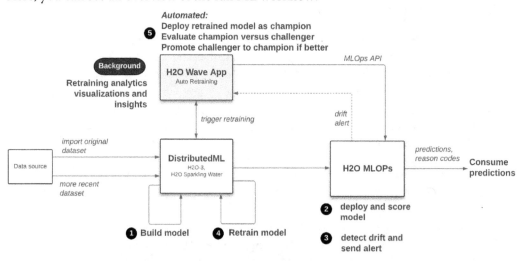

Figure 14.10 – A Wave app for automated model retraining

The workflow is summarized as follows:

1. The model is built and evaluated in H2O-3 or Sparkling Water (or other H2O model-building engines, such as Driverless AI).

2. The model is deployed to H2O MLOps and predictions are consumed. MLOps is configured to detect data drift on the model.

3. At a point in time, drift exceeds configured thresholds for the model, and an alert is sent to the model-retraining Wave app that you have built.

4. The model-retraining Wave app triggers the retraining of the model on the H2O model-building engine (in our case, H2O-3 or Sparkling Water). The Wave app deploys the retrained model as a challenger in H2O MLOps (using the MLOps API), and after a time, the Wave app evaluates the performances of the newly retrained challenger versus the existing champion model. The challenger is promoted to replace the champion if the former outperforms the latter. The cycle (*steps 3 and 4*) continues from this point.

The model-retraining Wave app can provide reporting, visualizations, and analytics around model retraining. For example, there could be a table of retraining history including time of retraining, drift measurement, current status (for example, training in progress, model deployed, and in challenger state or champion state), and so on. Visualizations could be provided that provide greater insights into the data drift and model-retraining and deployment pipeline. As an alternative, automation could be replaced by a human-in-the-loop workflow where steps in the pipeline are done manually based on data-scientist evaluations.

The goal of H2O Wave as an application-building framework is for you to easily build applications according to your own specifications and to integrate into the application other components in your ecosystem. So, you likely envision a model-retraining application a bit differently than what is shown here. The H2O Wave SDK allows you to build the application that you envision.

In our example model-retraining application, we integrated multiple H2O components into a Wave application workflow with visualizations and analytics. In the next section, we will expand integrations to non-H2O components of your ecosystem and thereby present a powerful framework to build Wave apps as a single pane of glass across your AI ecosystem.

A Reference H2O Wave app as an enterprise AI integration fabric

The low-code Wave SDK allows data scientists, ML engineers, and software developers to build applications that integrate one or more H2O components participating in the ML life cycle into a single application. H2O Wave is thus a powerful integration story.

Two Wave design facts need to be revisited, however, because they make this integration story even more powerful. First, Wave apps are deployed in containers and are thus isolated from other Wave apps. Second, developers can install and integrate publicly available or proprietary Python packages and APIs into the application. This means that H2O Wave apps can integrate both H2O and non-H2O components into a single application. This can effectively be restated as follows: H2O apps can be built as single panes of glass across your entire AI-related enterprise ecosystem. This is shown in the following diagram:

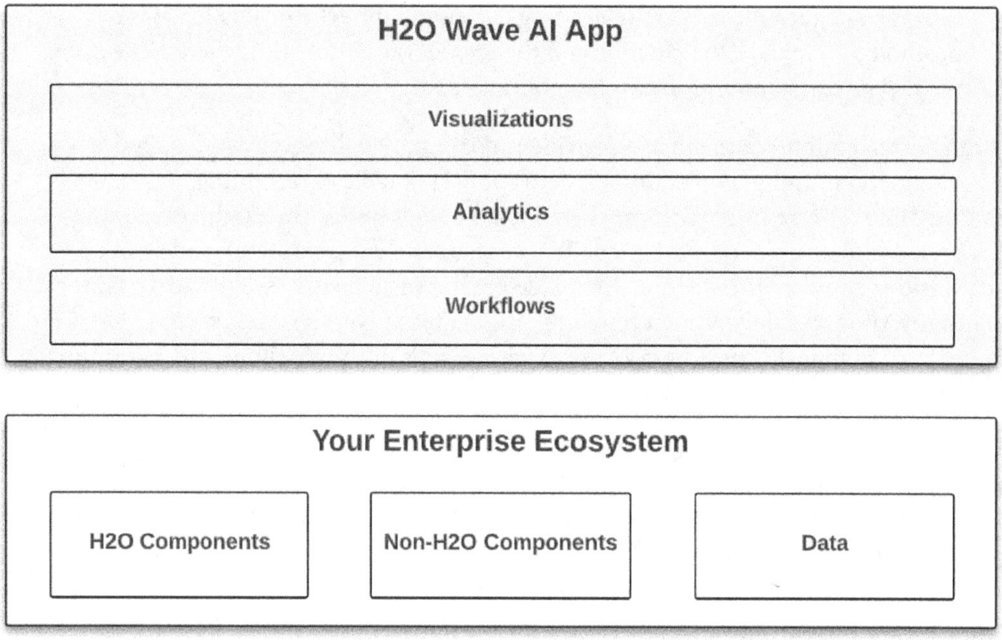

H2O Wave AI App

Visualizations

Analytics

Workflows

Your Enterprise Ecosystem

| H2O Components | Non-H2O Components | Data |

Figure 14.11 – H2O Wave AI app as a layer across your enterprise ecosystem

Data scientists, engineers, and software developers can thus build Wave apps that combine the end-to-end ML platform of H2O AI Cloud with AI-related and non-AI-related components of the enterprise ecosystem and its underlying cloud services. The diverse applications are hosted on H2O App Store with role-based access and thus made available to diverse enterprise stakeholder consumers. Let's break this down further by exploring the following diagram.

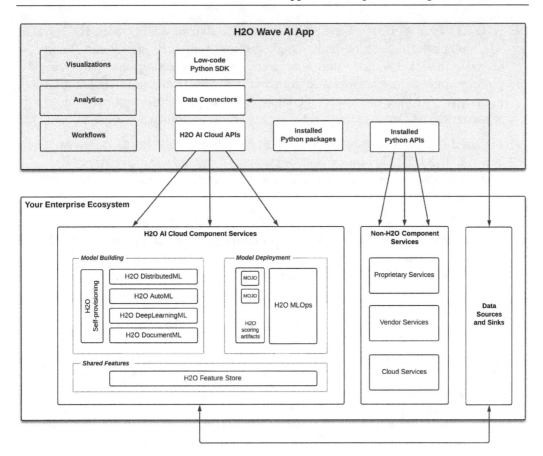

Figure 14.12 – Reference H2O Wave AI app layering across your enterprise ecosystem

This is a reference H2O Wave AI app showing its full potential to serve as a UI integration layer across your entire AI-related ecosystem. The goal is to show the full set of capabilities in this regard, and for you to use your imagination to instantiate this generalized reference into specific applications that fit your specific AI needs.

Let's review these capabilities, as follows:

- **Low-code Python SDK**: Wave's low-code Python SDK allows data scientists, ML engineers, and software developers to develop AI applications in a familiar Python style, focusing on populating data in widgets and templates and ignoring the complexity of web applications. The SDK can be extended with CSS style sheets to accomplish a specific look and feel if desired.

- **Data connectors**: The Wave SDK has over 140 connectors to diverse data sources and targets, making it easy to integrate Wave applications into your data ecosystem.

- **H2O AI Cloud APIs**: The Wave SDK has APIs for all components in the H2O AI Cloud platform: the four model-building engines and their provisioning tool, as well as the MLOps and Feature Store components. These integrations provide powerful ways to interact with all aspects of the ML life cycle from an application perspective. We have taken a quick glimpse of these possibilities in the scoring-consumption and model-retraining Wave applications discussed previously.

- **Installed Python packages**: The Wave SDK is extensible to any Python package you want to install. This allows you, for example, to extend the native Wave UI components with more specialized plotting or interactive visualization capabilities, or to use familiar packages to manipulate data.

- **Installed Python APIs**: You can also install Python libraries that serve as APIs to the rest of your enterprise ecosystem, whether it's your own components, non-H2O vendor components, or applications and native cloud services. This is a very powerful way to orchestrate ML workflows driven by APIs connected to H2O AI Cloud components with the capabilities across the rest of your enterprise ecosystem.

The capabilities and integrations just outlined open up a near unlimited number of ways for Wave apps to accomplish enterprise-AI analytics and workflows. You can, for example, integrate your model deployment and monitoring on H2O MLOps to existing multistep governance-process workflows. You can build UI workflows where users search data catalogs for authorized data sources, select a data source, and then launch and access an H2O model-building environment with the data source loaded. These are only two examples to get your mind started. As shown in *Figure 14.11*, there are many pieces you can tie together in your own specific and creative ways to extend ML beyond model building and scoring and to a larger context of workflows and multi-stakeholder business value.

Summary

In this chapter, we explored how H2O-at-scale technology (H2O-3, H2O Sparkling Water, H2O Enterprise Steam, and the H2O MOJO) expands its capabilities by participating in the larger H2O AI Cloud end-to-end machine learning ML platform. We saw, for example, how H2O-3 and Sparkling Water can gain from initial rapid prototyping and automated feature discovery. Likewise, we saw how H2O-3 and Sparkling Water models can be deployed easily to the H2O MLOps platform where they gain value from its model-scoring, monitoring, and management capabilities. We also saw how H2O AI Feature Store can operationalize features for sharing, both in model building with H2O-3 or Sparkling Water and model scoring on H2O MLOps.

We started exploring the power of H2O's open source low-code Wave SDK, and how data scientists, ML engineers, and software developers can use it to easily create visualizations, analytics, and workflows across H2O components and thus the full ML life cycle. These applications are published to the App Store component of the H2O platform where they are consumed by enterprise stakeholders or external partners or customers of the enterprise. One example Wave app that we explored was an employee-churn application to consume, understand, and respond to predictions on how likely individuals were to leave a company. Another was a model-retraining application where data scientists manage and track automated model-retraining workloads by leveraging the Wave app's underlying SDK integration with H2O-3 and H2O MLOps.

Finally, we introduced a reference Wave AI app to build applications that layer across H2O and non-H2O parts of the enterprise-AI ecosystem and thus form an enterprise-AI integration fabric.

We thus finish this book by taking ML at scale with H2O and putting it into the context of H2O's larger end to end machine learning ML platform called H2O AI Cloud. By marrying its established and proven H2O at scale technology with the new and rapidly innovating H2O AI Cloud platform, H2O.ai is continuing to prove itself as a bleeding-edge player in defining and creating new ML possibilities and value for the enterprise.

Appendix
Alternative Methods to Launch H2O Clusters

This appendix will show you how to launch H2O-3 and Sparkling Water clusters on your local machine so that you can run the code samples in this book. We will also show you how to launch H2O-3 clusters in the 90-day free trial environment for the H2O AI Cloud. This trial environment includes Enterprise Steam to launch and manage H2O clusters on Kubernetes infrastructure.

Note on Environments

Architecture: As introduced in *Chapter 2, Platform Components and Key Concepts*, you will use a client environment (with the H2O-3 or Sparkling Water libraries implemented) to run commands against a remote H2O-3 or Sparkling Water architecture distributed across multiple server nodes on a Kubernetes or Hadoop cluster. For small datasets, however, the architecture can be launched locally as a single process on the same machine as the client.

Versions: Functionality and code samples from this book use the following versions: H2O-3 version 3.34.0.7, and Sparkling Water version 3.34.0.7-1-3.2 to run on Spark 3.2. You will set up your environment with the latest (most recent) stable versions, which will allow you to run the same code samples from this book but will also include capabilities in H2O-3 and Sparkling Water that were added after the book was written.

Languages: You can set up your client environment in Python, R, or Java/Scala. We will use Python in this book. Your Python client can be a Jupyter notebook, PyCharm, or other.

Let's learn how to run H2O-3 entirely in your local environment.

Local H2O-3 cluster

This is the easiest method to run H2O-3 and is suitable for the small datasets used in code samples in this book. It launches H2O-3 on your local machine (versus an enterprise cluster environment) and does not involve H2O Enterprise Steam.

First, we will perform a one-time setup of our H2O-3 Python environment.

Step 1 – Install H2O-3 in Python

To set up your H2O-3 Python client, simply install three module dependencies in your Python environment and then the h2o-3 Python module. You must use Python 2.7.x, 3.5.x, 3.6.x, or 3.7.x.

More specifically, do the following:

1. Install dependencies in your Python environment:

    ```
    pip install requests
    pip install tabulate
    pip install future
    ```

2. Install the H2O-3 library in your Python environment:

```
pip install h2o
```

Please refer to `http://h2o-release.s3.amazonaws.com/h2o/rel-zumbo/1/index.html` (the **INSTALL IN PYTHON** tab) to install H2O-3 in Conda.

You are now ready to run H2O-3 locally. Let's see how to do that.

Step 2 – Launch your H2O-3 cluster and write code

To start a local single-node H2O-3 cluster, simply run the following in your Python IDE:

```
import h2o
h2o.init()
# write h2o-3 code, including code samples in this book
```

You can now write your H2O-3 code, including all samples from this book. See *Chapter 2, Platform Components and Key Concepts,* for a `Hello World` code sample and an explanation of what happens under the surface.

Java Dependency – Only When Running Locally

The H2O-3 cluster (not the Python client) runs on Java. Because you are running the cluster on your local machine here (representing a single-node cluster), you must have Java installed. This is not required when you use your Python client to connect to a remote H2O cluster in your enterprise Kubernetes or Hadoop environment.

Now, let's see how we can set up our environment to write Sparkling Water code on our local machine.

Local Sparkling Water cluster

Running Sparkling Water locally is similar to running H2O-3 locally, but with Spark dependencies. See this link for a full explanation of the Spark, Python, and H2O components involved: `https://docs.h2o.ai/sparkling-water/3.2/latest-stable/doc/pysparkling.html`.

We will be using Spark 3.2 here. To use a different version of Spark, go to the **Sparkling Water** section of the H2O downloads page at the following link: `https://h2o.ai/resources/download/`.

For your Sparkling Water Python client, you must use Python 2.7.x, 3.5.x, 3.6.x, or 3.7.x. We will be running Sparkling Water from a Jupyter notebook here.

Step 1 – Install Spark locally

Follow these steps to install Spark locally:

1. Go to `https://spark.apache.org/downloads.html` to download Spark. Make the following choices and then download:

    - Spark version: 3.2.x

    - Package type: Pre-built for Hadoop 3.3 and later

2. Unzip the downloaded file.

3. Set the following environment variables (shown here for macOS):

    ```
    export SPARK_HOME="/path/to/spark/folder"
    export MASTER="local[*]"
    ```

Now, let's install the Sparkling Water library in our Python environment.

Step 2 – Install Sparkling Water in Python

Install the following modules:

1. Install dependencies in your Python environment:

    ```
    pip install requests
    pip install tabulate
    pip install future
    ```

2. Install the Sparkling Water Python module (called `PySparkling`). Note the module reference to Spark 3.2 specifically here:

    ```
    pip install h2o_pysparkling_3.2
    ```

Next, let's install an interactive shell.

Step 3 – Install a Sparkling Water Python interactive shell

To run Sparkling Water locally, we need to install an interactive shell to launch the Sparkling Water cluster on Spark. (This is only required when running Sparkling Water locally; Enterprise Steam takes care of this when running on your enterprise cluster.) To do so, perform the following steps:

1. Download the interactive shell by navigating to the **Sparkling Water** section of `https://h2o.ai/resources/download/`, clicking **Sparkling Water For Spark 3.2**, and finally, clicking on the **DOWNLOAD SPARKLING WATER** button.

2. Unzip the download.

Now, let's launch a Sparkling Water cluster and access it from a Jupyter notebook.

Step 4 – Launch a Jupyter notebook on top of the Sparkling Water shell

We assume you have Jupyter Notebook installed in the same Python environment as your installations in step 2. Perform the following steps to launch a Jupyter notebook:

1. On the command line, navigate into the directory where you unzipped the download in step 3 of this section.

2. Launch the Sparkling Water interactive shell and a Jupyter notebook in it:

   - For macOS, use the following:

   ```
   PYSPARK_DRIVER_PYTHON="ipython" \
   PYSPARK_DRIVER_PYTHON_OPTS="notebook" \
   bin/pysparkling
   ```

   - For Windows, use the following:

   ```
   SET PYSPARK_DRIVER_PYTHON=ipython
   SET PYSPARK_DRIVER_PYTHON_OPTS=notebook
   bin/pysparkling
   ```

   Your Jupyter notebook should launch in your browser.

Now, let's write Sparkling Water code.

Step 5 – Launch your Sparkling Water cluster and write code

In your Jupyter notebook, type the following code to get you started:

1. Start your Sparkling Water cluster:

```
from pysparkling import *
import h2o
hc = H2OContext.getOrCreate()
hc
```

2. Test the installation:

```
localdata = "/path/to/my/csv"
mysparkdata = spark.read.load(localdata, format="csv")
myH2Odata = hc.asH2OFrame(mysparkdata)
```

You are now ready to build models using both H2O and Spark code.

H2O-3 cluster in the 90-day free trial environment for H2O AI Cloud

Here, you must interact with Enterprise Steam to run H2O-3. In this case, you will install the h2osteam module in your Python client environment in addition to the h2o module as we did when running H2O-3 locally.

Step 1 – Get your 90-day trial to H2O AI Cloud

Get your trial access to H2O AI Cloud here: https://h2o.ai/freetrial.

When you have completed all steps and can log in to H2O AI Cloud, then we can start running H2O-3 clusters as part of the H2O AI Cloud platform. Here are the next steps.

Step 2 – Set up your Python environment

To set up your Python client environment, perform the following steps:

1. Log in to H2O AI Cloud and click on the **My AI Engines** tab. This will take you to Enterprise Steam, as shown in the following screenshot. From there, download the h2osteam library by clicking on the **Python Client** option from the sidebar:

Figure 15.1 – Enterprise Steam

2. Install the h2osteam library in your Python environment by running the
    following command:

    ```
    pip install /path/to/download.whl
    ```

    Here, /path/to/download.whl is replaced by your actual path.

3. You will also need to install the h2o library. To do so, execute the following:

    ```
    pip install requests
    pip install tabulate
    pip install future
    pip install h2o
    ```

Now, let's use Steam to start an H2O cluster and then write H2O code in Python.

Step 3 – Launch your cluster

Follow these steps to launch your H2O cluster, which is done on a Kubernetes server cluster:

1. In Enterprise Steam, click **H2O** on the sidebar and then click the **Launch
    New Cluster** button.

2. You now can configure your H2O cluster and give it a name. Be sure to configure
    the latest H2O version from the dropdown, which should match the library you
    installed in the previous step.

3. When configured, click the **Launch Cluster** button and wait for the cluster launch to complete.

4. You will need the URL to Enterprise Steam to connect to it from your Jupyter notebook or other Python client. While in Steam, copy the URL from `https` to `h2o.ai`, inclusive.

Step 4 – Write H2O-3 code

We can now start writing code (for example in Jupyter) to build models on our H2O-3 cluster that we just launched. Perform the following steps after opening your Python client:

1. Import your libraries and connect to Enterprise Steam:

```
import h2o
import h2osteam
from h2osteam.clients import H2oKubernetesClient

conn = h2osteam.login(
    url="https://SteamURL,
verify_ssl=False,
    username="yourH2OAICloudUserName",
    password=" yourH2OAICloudPassword")
```

> **Important Note**
>
> At the time of this writing the URL for the 90-day H2O AI Cloud trial is `https://steam.cloud.h2o.ai`.
>
> For password you can use your login password to the H2O AI Cloud trial environment, or you can use a temporary personal access token generated from the Enterprise Steam Configurations page.

2. Connect to your H2O cluster you started in Enterprise Steam:

```
cluster = H2oKubernetesClient().get_cluster(
    name="yourClusterName",
    created_by="yourH2OAICloudUserName")

cluster.connect()
# you are now ready to write code to run on this H2O
cluster
```

You can now write your H2O-3 code, including all samples from this book.

Index

A

N

O

P

R

Packt.com

Subscribe to our online digital library for full access to over 7,000 books and videos, as well as industry leading tools to help you plan your personal development and advance your career. For more information, please visit our website.

Why subscribe?

- Spend less time learning and more time coding with practical eBooks and Videos from over 4,000 industry professionals

- Improve your learning with Skill Plans built especially for you

- Get a free eBook or video every month

- Fully searchable for easy access to vital information

- Copy and paste, print, and bookmark content

Did you know that Packt offers eBook versions of every book published, with PDF and ePub files available? You can upgrade to the eBook version at packt.com and as a print book customer, you are entitled to a discount on the eBook copy. Get in touch with us at customercare@packtpub.com for more details.

At www.packt.com, you can also read a collection of free technical articles, sign up for a range of free newsletters, and receive exclusive discounts and offers on Packt books and eBooks.

Other Books You May Enjoy

If you enjoyed this book, you may be interested in these other books by Packt:

Learn Amazon SageMaker, Second Edition

Julien Simon

ISBN: 9781801817950

- Become well-versed with data annotation and preparation techniques
- Use AutoML features to build and train machine learning models with AutoPilot
- Create models using built-in algorithms and frameworks and your own code
- Train computer vision and natural language processing (NLP) models using real-world examples
- Cover training techniques for scaling, model optimization, model debugging, and cost optimization
- Automate deployment tasks in a variety of configurations using SDK and several automation tools

Automated Machine Learning on AWS

Trenton Potgieter

ISBN: 9781801811828

- Employ SageMaker Autopilot and Amazon SageMaker SDK to automate the machine learning process
- Understand how to use AutoGluon to automate complicated model building tasks
- Use the AWS CDK to codify the machine learning process
- Create, deploy, and rebuild a CI/CD pipeline on AWS
- Build an ML workflow using AWS Step Functions and the Data Science SDK
- Leverage the Amazon SageMaker Feature Store to automate the machine learning software development life cycle (MLSDLC)
- Discover how to use Amazon MWAA for a data-centric ML process

Packt is searching for authors like you

If you're interested in becoming an author for Packt, please visit `authors.packtpub.com` and apply today. We have worked with thousands of developers and tech professionals, just like you, to help them share their insight with the global tech community. You can make a general application, apply for a specific hot topic that we are recruiting an author for, or submit your own idea.

Share Your Thoughts

Now you've finished *Machine Learning at Scale with H2O*, we'd love to hear your thoughts! Scan the QR code below to go straight to the Amazon review page for this book and share your feedback or leave a review on the site that you purchased it from.

`https://packt.link/r/1800566018`

Your review is important to us and the tech community and will help us make sure we're delivering excellent quality content.

www.ingramcontent.com/pod-product-compliance
Lightning Source LLC
LaVergne TN
LVHW081330050326
832903LV00024B/1100